DATE DUE

DE 18 '99			
AG 5 '00			
NO 28 '00			
NO 21 '00			
JE 14 '01			
DE 19 '01			
DE 21 '02			

DEMCO 38-296

The Genetic Gods

The Genetic Gods

Evolution and Belief in Human Affairs

John C. Avise

Harvard University Press

Cambridge, Massachusetts and London, England, 1998

Permission has been granted by the publisher to quote from *Old Turtle* by Douglas Wood
(Duluth, Minn.: Pfeifer-Hamilton Publishers, 1992).

Library of Congress Cataloging-in-Publication Data

Avise, John C.
    The genetic gods : evolution and belief in human affairs / John C. Avise.
        p.   cm.
    Includes bibliographical references and index.
    ISBN 0-674-34625-4 (cloth : alk. paper)
    1. Genetics—Popular works.
    2. Evolutionary genetics—Popular works.
    I. Title.
    QH437.A95   1998
    576—dc21   98-3255

Contents

Preface

This book is intended for the reflective, open-minded reader who would appreciate a simplified discussion of recent evolutionary-genetic findings. Human beings, like all other species on earth, are biological products of evolutionary processes, and as such are physical expressions of genes, the "genetic gods." Genes and the mechanistic evolutionary forces that have sculpted them thus assume many of the roles in human affairs traditionally reserved for supernatural deities. Some may find this argument blasphemous or sacrilegious; others may find it prosaic. Such contradictory responses reflect the paradoxical state of philosophical affairs, in which religious revelation and scientific rationalism uncomfortably coexist as powerful but opposing means of knowing.

During the development of an individual, genes influence not only bodily features at microscopic and macroscopic levels and the metabolic and physiologic conditions underlying medical health, but also the more ethereal aspects of human nature, including emotions, psychologies, personalities, and even ethical and religious predilections. These genetic influences often are indirect, mediated and modulated by diverse social and cultural experiences, and manifest through genetically based cognitive abilities unique to our species. Even the most sacrosanct of human affairs, sexual reproduction and death, are products of evolutionary processes, firmly ensconced in our genes.

The sciences of evolutionary biology and genetics, with roots little more than a century deep, have blossomed in recent decades to provide mechanistic understandings of human conditions that until recently had been within the exclusive purview of mythology, theology, and religion. Yet most people remain either blissfully ignorant of these discoveries or openly hostile to their implications. As a practicing evolutionary biologist, I live in two worlds. I work in a university setting surrounded by the astonishing pieces of laboratory equipment and biochemical tools of molecular genetics, and by reams of computer output on DNA sequences and evolutionary-genetic simulations. My colleagues and graduate students take for granted that natural forces have shaped the biological objects of our studies (in my case, mostly fishes, reptiles, and birds), and research grants fund us to work out the genetic mechanisms and processes by which these evolutionary outcomes have been achieved. Yet when I return home to read the local newspaper, I find editorials lashing out against evolutionary biology in the name of religion, and reports of school boards mandating equal time for creationism in the science classroom. On TV and radio, evangelists prohibit any departure from the word of God as they hear it. Almost as disturbing are the technocrats or laboratory researchers who naively proclaim that science and technology alone can provide certain salvation from humanity's ills and all the world's problems.

This is a book about causation in biology. It makes no pretense to wrestle seriously with the theistic ramifications of evolution from the perspectives of religious philosophers or theologians, who also have dealt with such issues extensively.[1] However, a clearer understanding of recent empirical findings in human molecular genetics and conceptual advances in evolutionary-genetic theory may increase communication between the social and the natural sciences, and between theology and evolutionary biology. I hope to diminish the hostility between these differing epistemological approaches, which at their best do share a goal of attempting to understand human nature.

Beyond these immediate aims, I hope to resolve a central issue in my own life: how to reconcile the intellectual demands and pleasures of critical scientific thought with the sense of purpose and fulfillment that a rich spiritual life can provide. I subscribe to the proposition that scientific rationalism is the surest route to objective understanding available to mortal humans, but I hold no illusions that the pursuit of objective reality is necessarily satisfying. This book champions science as the preferred path of rational inquiry, but it makes no definitive claims regarding either the ethical or the pragmatic value of rational objectivity itself.

I wish to thank Francisco Ayala, Betty Jean Craig, Douglas Futuyma, Mike Goodisman, Matt Hare, Glenn Johns, Adam Jones, Bill Nelson, Guillermo Ortí, Paulo Prodöhl, Daniel Promislow, DeEtte Walker, Janet Westpheling, Kurt Wollenberg, and several anonymous reviewers for critical comments on various drafts of this manuscript. Through their conscientious efforts, I have gained a deeper appreciation for the great diversity of thoughtful opinions regarding evolutionary causation. I would also like to thank my editors Michael Fisher and Kate Brick at Harvard University Press for encouragement and excellent support.

Prologue

In one of my favorite Star Trek episodes, the crew of the USS Enterprise is sent to intercept an unknown object rapidly approaching Earth from deep space. As the story unfolds, the entity reveals itself as "VGER" (pronounced *veejer*), and announces its intent: to find its creator on the third planet of our solar system. VGER proves to be the long-lost *Voyager* 6 spacecraft, launched from Earth three centuries earlier, and programmed to explore the universe. Despite having amassed vast scientific knowledge on its fact-finding journey, VGER's computer systems nonetheless remain unfulfilled, and the space module now feels compelled to discover its maker. The unsuspecting VGER is aghast and chagrined to learn that mere humans ("carbon-based units") are its gods.

Of course, VGER had discovered merely its proximate makers. Humans must have their own creator. For thousands of years, we too have been in search of deities through the emotional experiences and exercises of faith that characterize mythologies and traditional religious practices. Only in the last two centuries have we begun an alternative, scientific approach to the exploration of our biotic origins. Like the Star Trek Vulcans, whose emotionless logic contrasts with the cultivated ardor of many religious practitioners and theologians, scientists operate under the principle that explanations for natural phenomena should be sought using observations and hypotheses divorced from emotive appeals, subjective or personal impressions, or untestable dogma. Science and blind faith are essentially at odds. Yet, like StarTrek's Spock (whose father was a Vulcan, mother a human), individual scientists find it impossible in practice to divorce their thoughts and findings from social context, or from emotional or religious experiences, and I'm quite certain most of them wouldn't wish to do so.

1

The scientific search for our creators frequently has clashed with more traditional ways of interpreting our origins. Science is only one avenue to understanding the world and universe. Nonetheless, it is a new and revolutionary method in human history that differs fundamentally in mode of inquiry from all prior approaches. Just as VGER discovered its gods, in the last century scientists have discovered our proximate makers. They are our genes. We now know in detail the origin of these genes, and understand the evolutionary forces that have shaped them, and hence ourselves. We and all other species on earth serve our respective genetic materials, which in turn have been altered by mutation and recombination, and contingently shaped by natural selection. These findings have had and will continue to have a profound impact on the structure of human belief systems.

ONE

The Doctrines of Biological Science

They mastermind our lives, influencing our physical appearance, health, behavior, even our fears and aspirations. They constitute our material reason for being—for eating and sleeping, warring and loving, hating and caring, forging relationships—for procreation. To them we owe our existence; on us they depend for continuance. We are their evolutionary inventions, ephemeral and disposable hosts. We have been sculpted unwittingly into who we are by serving their replicative needs, but they too have been shaped by this servitude in a continuing cycle of reproductive reciprocity so fundamental as to blur the distinction between them and us. We are their ticket to immortality, but their enthusiasm for our physical well-being fades and finally expires as we pass reproductive age. They give us life, yet dictate senescence and death. "They" are not gods, but our genes.

Describing genes as "gods," however, is particularly apt. First, genes have special powers over human lives and affairs, as anyone stricken with a serious genetic disorder will attest. Second, genes exert influence over the course of nature. Indeed, the course of nature *is* biological evolution itself, which can be defined in simplest terms as genetic change through time. Third, gene lineages are potentially immortal, and demonstrably so over billion-year time scales. The genetic material in organisms alive today traces back generation by generation through an unbroken chain of descent (with modification) from ancestral molecules that have copied and replaced themselves ever since the origin of life on earth, about 4 billion years ago. In complex multicellular organisms such as vertebrate animals, these surviving genes have been transmitted through germ cells (eggs and sperm), rather than through the somatic cells that make up brains, hearts, muscles, livers, eyes, and all other tissues, organs, and body parts. Only germline genes are potentially immortal: somatic entities (ourselves included) are merely ephemeral vessels that evolved as a means of perpetuating DNA.

Perhaps we register a general awareness of the power of genes in our gut responses to the prospects of human genetic cloning. When scientists recently cloned a sheep from the genes of an adult ewe, lawmakers in several countries immediately called for bans on similar experimentation with humans; theologians pondered whether a cloned human body would carry a cloned soul as well; and many monitored such developments with great curiosity and concern. If genes were merely passive backdrops to human existence, why should we care about the prospects of human cloning? But most of us do care. We understand, if only at an impressionistic level, the concept of genetic individuality, and of the hereditary influences that help to make you, you and me, me. We nod politely (and properly) to obligatory reminders about the importance of environmental influences on human behavior and character. In places of worship, we pay homage to supernatural gods and their supposed overriding jurisdiction over human affairs. Yet we also correctly appreciate the pervasive authority of genes.

Since genes, unlike gods, are natural entities (see Figure 1.1), their origins and designs are suitable subjects for scientific inquiry, and can be understood in terms of physicochemical mechanisms and evolutionary processes without recourse to occult, mystical, or magical explanations. Natural phenomena can be awe-inspiring, but explanations for these phenomena that invoke miracles in the sense of magic fall outside the realm of science. Unlike supernatural deities, the genetic gods can be studied objectively and analyzed critically.

All successful biological research in the twentieth century has involved three doctrines: mechanism, natural selection, and historicity.[1]

The doctrine of mechanism suggests that all functions and processes in living organisms in principle are understandable in terms of physical and chemical phenomena, as played out in an evolutionary theatre. Mechanism is rather new as an orienting philosophy for the study of life. Theologians as well as biologists of earlier centuries commonly assumed that vitalistic or supernatural forces purposefully directed life toward goals preordained by one or more

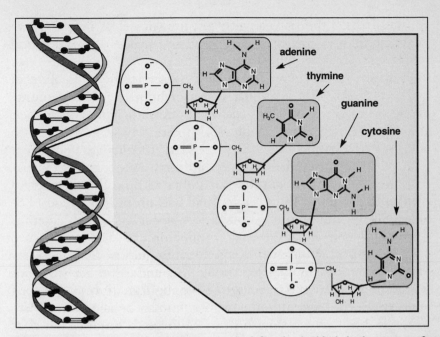

Figure 1.1 The genetic material, DNA. On the left is the double-helical structure of a short stretch of DNA. The ribbons that constitute the backbone of the molecule are alternating phosphate groups and sugars, from which project inward the four types of chemical bases [adenine (A), thymine (T), guanine (G), or cytosine (C)] that distinguish DNA's nucleotide building blocks. On the right is a closeup of the molecular structures of the four nucleotides, each of which is composed of a deoxyribose sugar connected to a phosphate group (open circle) and to a particular base (darkened box). A DNA molecule consists of long, complementary, double-helical stretches of these nucleotides, in which the paired strands of the double helix are held together by hydrogen bonds between A's and T's, and between G's and C's. The genome in each human cell consists of three billion pairs of these nucleotides in ordered strings. A typical gene is about 20,000 nucleotide pairs in length.

omnipotent creators. Vitalism and teleology (the antitheses of mechanism) remain at the heart of many formal religions, which tend to assume strict dualities of body and mind, body and soul, and of physical presence from the hereafter. In such religions, nonmaterialistic souls, moralities, angels, devils, heavens, hells, and ultimate purpose are thought to exist and to operate outside the rules of biological mechanism.

However, the biological sciences provisionally shelve vitalist theories behind the alternative hypothesis that living systems operate according to mechanistic processes subject to critical and objective evaluation. These mechanistic processes in science must be verifiable by independent observers, show logical and empirical self-consistency, display similar consistency with related external events or processes, and be capable of generating predictions that at least potentially are falsifiable. Only after passing these rigorous tests can a scientific idea be accepted as a provisional truth. If biological systems can be described and understood in a mechanistic sense, then for the purposes of understanding life there is no compelling scientific rationale to erect additional, teleological explanations for the origin and operation of living systems. Alternatively, if uncritical teleological or supernatural explanations are adopted a priori, all impetus for rational inquiry is lost.

No doubt this mutual exclusivity is one reason for the cool reception afforded evolutionary biology by some formal religions (and vice versa). A common perception is that as scientific understanding grows, the universe of phenomena interpretable solely under the auspices of religious faith shrinks. We might think instead of contemporary scientific research as a balloon. The air inside the balloon would be the sum of our mechanistic knowledge. As the volume of the balloon expands through scientific discovery, so too does its surface area, its interface with the unknown, exposing ever-enlarging horizons for further inquiry. An entrepreneurial antiscientist always can speculate about what lies beyond such horizons, and committed teleologists always can entertain hope that the entire scientific balloon one day will burst as it encounters some prickly biological phenomenon truly outside the realm of mechanism. In any event, a critical assumption of this metaphor is that the universe of the unknown is far larger than the universe of the known, a proposition on which scientists and theologians perhaps can agree.

There is a related sense in which scientific discoveries can expand the horizons and jurisdictions of theology. Consider, for example, the current revolution taking place in recombinant DNA technology, and its many possibilities for human genetic engineering. As emphasized repeatedly in this book, scientific findings are amoral: they may help to explain the operations of nature, but they alone cannot dictate how things "should" be on any ethical scale. Our ability to test, screen, and manipulate the human genome is made possible by recent technological breakthroughs of science, but whether and how we might exercise such unprecedented powers must be considered by, among others, ethicists and theologians. No religions of past centuries were presented with such issues. Indeed, before the middle of the nineteenth century, no religions were afforded the opportunity to reconsider human nature in the light of findings from the evolutionary and genetic sciences.

In accounting for life's operations, one possibility technically consistent with mechanism is that there truly is (or was)[2] an omnipotent god with a grand design for life, but that this noninterventionist deity long ago set into motion and has permitted the continued operation of scientifically intelligible mechanistic processes. If so, a wise bet might be on scientists (rather than shamans, priests, rabbis, or Sufis) to be the first to achieve an *objective* understanding of this god.[3]

Within the last 150 years, biologists have banished much of the mystery surrounding several "geneses" that dominated theological and religious discourse for millennia: the development of an individual organism (from a fertilized egg); the dawn of the human species (from a common ancestor with the great apes); and the beginning of all self-perpetuating life (from inorganic materials in a primordial earth). Imagine that a traditional religion had demonstrated convincingly, rather than merely asserted, the nature of these awesome happenings. Surely that would be a powerful religion, one worthy of widespread attention and admiration. Yet science has not always received such respect, and many continue to operate as though science were an adversary rather than an ally in attempts to decipher the context of our existence.

Detractors of science often emphasize that there is much that science cannot yet explain, as if this were somehow a fatal blow to the endeavor. Scientific understanding will always be incomplete, since the boundaries of knowledge are forever expanding. Good science normally generates additional questions. To pick one obvious example, scientists still have only the most rudimentary understanding of the mechanistic workings of the human brain.[4]

However, we have learned more about the brain in the last one hundred years through the scientific method than in all previous centuries combined, and the pace of learning continues to accelerate.

The field of evolutionary genetics is based on two independent discoveries published within a short span of time in the middle of the nineteenth century. In 1859, a former shipboard naturalist produced the most important and influential book in the history of biology, *On the Origin of Species*. By identifying natural selection as a primary agent of evolutionary change, Charles Darwin forever altered our view of life by demonstrating that, over many generations, natural forces can account for adaptations within a species that otherwise would seem to be the work of a sentient creator. Six years later, in 1865, a solitary monk working in a small monastery in what is now the Czech Republic published a technical paper on transmittable factors in pea plants.[5] By identifying a previously unsuspected mode of "particulate" transmission for hereditary factors (not to be named genes until 1909), Gregor Mendel fundamentally altered all prior notions about the operation of biological inheritance. These founding fathers of evolutionary biology and of genetics, respectively, had uncovered two of the most fundamental themes of life, setting the agenda for nearly all biological research conducted since. This scientific agenda was not biblically mandated. No biologist could have a fonder professional dream than to make an astounding discovery that would overturn or dramatically improve the now-conventional wisdoms of Darwinian natural selection or Mendelian heredity. Rather, the research agenda of the last century developed because, despite exceptional effort, no other scientific explanations as yet have proven nearly so powerful in deciphering life's properties.

The works of Darwin and Mendel exemplify the great variety of routes to fundamental scientific advance. Darwin's treatise was a compositionist work, the inductive elaboration of a conceptual insight distilled from life-long observations of the natural world. His book was a best-seller in its day, read by both scientists and the general public. Its appeal stemmed from the compelling logic and empirical strength of the argument for natural selection, as well as the audacity of its implications.[6] By contrast, Mendel's work was a reductionist, deductive documentation of an unanticipated mode of heredity, based on controlled experimentation and detailed numerical analyses of progeny arrays in crosses between strains of pea plants. Mendel's findings went unnoticed at the time, only to be discovered by the scientific community in 1900, sixteen years after his death.

Natural selection is the second doctrinal foundation of twentieth-century biological inquiry. No longer was the direct hand of a god needed to explain the otherwise miraculous match of species to their respective environments, nor to account for the marvelous morphological, physiological, and behavioral adaptations that organisms need to reproduce and survive.

It is little wonder that most theologians as well as natural historians before Darwin's time had been misled into creationist scenarios:[7] the adaptations produced by natural selection gave every indication of having been custom built by a supreme intelligence. But following the elucidation of natural selection, the products of evolution could be seen to arise entirely from natural causes, ending the agonizing question of how an all-caring and all-powerful god could permit evil and suffering in the world. With the recognition that evolution occurs under the direction of an indifferent mechanistic process, natural selection, this pivotal moral enigma evaporated, although for some a deeper question remained: Why would an omnipotent god permit heartless processes to guide life?

Natural selection is the simplest yet most difficult of concepts. Its essence is that some individuals and their genes tend to survive and reproduce better than others in a given environment, and these become disproportionately represented in subsequent generations. These superior reproducers are said to be more fit, and normally will be the types better adapted to existing environmental conditions. Yet there are subtleties in the working of natural selection (and its interaction with other evolutionary processes such as mutation and recombination) that make its study particularly challenging. As D. L. Hull notes, "Evolutionary theory seems so easy that almost anyone can misunderstand it."[8] Although evolutionary biology and genetics are among the most intellectually challenging and sophisticated arenas of scientific inquiry, their general accessibility to nonscientists often leads to the proffering of misinformation or pseudoscientific leaps of faith.

A brief list of some of the demanding technical and conceptual questions posed by the theory of natural selection demonstrates the breadth of its implications. On what level—genes, individuals, kinship groups, populations, species, higher taxa—does natural selection act most effectively? What are the domains, levels, patterns, and rates of natural selection in forging adaptations? What are the trade-offs among potentially competing components of fitness, such as survival and reproduction? Some questions concerning natural selection grade into areas of philosophy. Is natural selection rational, progressive, or tautological? Why hasn't natural selection achieved more perfect species through time (or has it)? Many such issues are of special relevance to the human condition. Does natural selection continue to operate in modern societies? Where might it lead? Can and should we intervene to prevent sickness or even death? These are but a few of the classes of questions that have challenged evolutionary biologists (and philosophers) ever since Darwin.

So ineluctable is natural selection, given the appropriate conditions, that the process can be expected to operate anywhere in the universe that life exists. Indeed, "natural" selection can be shown to operate in appropriately contrived nonliving systems. In the 1960s, John Holland at the University of Michigan introduced a genetic-algorithm approach to computer programming that mimics the problem-solving operation of evolution. In this method of machine learning, computer software is designed such that lines of code operate like living organisms under Darwinian selection. The algorithmic codes are allowed to mutate randomly, mate with other bits of code to form new combinations, compete against one another in obtaining desired solutions to engineering problems, and differentially reproduce according to this performance criterion: "survival of the cyberfittest." Without further human intervention, the software tends to bootstrap itself from poorer to better solutions, sometimes generating novel designs that outperform the blueprints of conventional conscious engineering. Given that an analogue of natural selection can operate in appropriately contrived nonliving systems, one compelling question for living systems then becomes: How does an appropriate milieu of conditions suitable for natural selection arise? In other words, how is the genetic variety generated that provides the fodder for natural selection, and how have these variability-generating processes themselves evolved? Such fundamental questions are at the heart of the science of evolutionary genetics.

Among the features that tend to distinguish life from nonlife, such as growth, metabolism, movement, and responsiveness to stimuli, the most fundamental is self-reproduction. All known forms of life possess this potential, usually at multiple organizational levels. DNA molecules replicate within cell lineages, somatic cells differentiate in the production of multicellular organisms, organisms replace themselves with progeny, and new species arise from ancestral forms during the speciation process. At any hierarchical level in such a system of self-proliferating elements, some categories of variants are likely to be better suited for survival and reproduction than others in certain environments, and thereby will tend to show proportionately increased representation at a later time. An opportunity thus arises for evolution by natural selection.

For this opportunity to be realized, the self-replications must be highly faithful, yet imperfect. Imagine a chaotic form of replication wherein progeny bear no greater resemblance to their genealogical ancestors than to unrelated reproducers (in other words, that heredity as we know it did not exist). There would be then no tendency for the attributes of reproductively favored classes to be disproportionately represented in successive generations, and, thus, no adaptive evolution. At the other end of the spectrum, imagine replication that is forever perfect. In the absence of variation among replicators, evolutionary change quickly would cease. In truth, however, all biological replications tend to be loyal yet imperfect. During DNA replication, errors (mutations) arise with low but detectable frequencies reflective of the molecules' chemical properties and of imperfections in genetically-based DNA repair processes. In organismal reproduction through sexual means, offspring tend to resemble their parents in particular features, yet differ genetically from them and from one another because of the gene-shuffling (recombinational) mechanisms of meiosis and syngamy. Meiosis is the cellular process by which eggs and sperm are produced, and syngamy is the union of these eggs and sperm (gametes) during fertilization (see Figure 1.2). On a larger scale, when members of one species diverge in a process called speciation, genetic variability in an ancestral taxon is partially converted to genetic differences between the similar daughter species that emerge.

One of the most challenging questions for contemporary scientific research is why imperfections in the reproductive process appear so well-suited for continued evolution by natural selection. For example, once a reasonably high level of adaptation has been achieved, it would seem that the short-term selective advantage might lie with perfect self-replicators, particularly in stable environments. Long-term evolution is contingent upon variation among replicating units, but natural selection has no known mode of operation by which it can plan for future needs. One scientific possibility is that short-term fitness advantages frequently stem from mutationally- and recombinationally-derived genetic variability such that these processes have been (and are) directly favored by natural selection. A competing hypothesis is that variation-generating processes tend to be disadvantageous in the short term but cannot be eliminated. For example, with respect to genetic recombination that is characteristic of sexual reproduction, histori-

Figure 1.2 Simplified diagram of the particulate nature of inheritance as discovered by Gregor Mendel, showing two pairs of chromosomes in a male and a female. On one of the pairs of chromosomes in these diploid parents occurs a gene that is shown as present in either of two allelic forms, "A" or "a". On the other pair of chromosomes is another gene, also with two alleles, "B" and "b". According to Mendel's first rule (the law of segregation), during the cellular process of meiosis that occurs in germline cells and produces haploid eggs and sperm, the two alleles of a gene are distributed at random to these gametes. Thus, for example, in this case about half of the gametes carry the "A" allele, and the others carry the "a" allele. Under Mendel's second rule (the law of independent assortment), during meiosis the alleles at separate genes assort independently, such that in this case four gametic types (AB, Ab, aB, and ab) are expected in about equal frequency. In the production of the next generation of diploid progeny, these gametes can unite to generate a plethora of genotypic combinations (particularly when large numbers of genes are considered). However, the alleles of each gene retain their separate identities during the sexual reproduction process. Hence, a "particulate" mode of inheritance characterizes all life, from pea plants to humans.

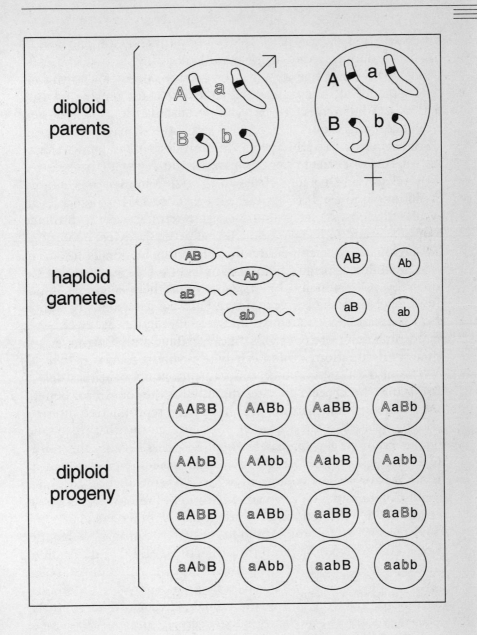

diploid parents

haploid gametes

diploid progeny

cal inertia and phylogenetic legacy may make it difficult for an organism to revert to asexuality regardless of any short-term fitness benefits. Perhaps recombinational variation arose as a fortuitous byproduct of cellular processes that evolved for repair of DNA damages. With respect to *de novo* mutations, the great size and sequence complexities of organismal genomes may preclude an absolutely error-free DNA replication process. All such possibilities are subjects of current scientific inquiry.

A related and equally challenging question in contemporary evolutionary research is: At what level, from DNA molecules to species, does natural selection operate most effectively to produce adaptations and to generate organismal diversity? Most evolutionary biologists now appreciate that natural selection is far more influential in forging adaptations at lower echelons in the biological hierarchy. Organismal adaptations benefit their encoding genes first, rather than the species, higher taxa, or ecosystems of which those units are a part.[9] Adaptations occur more often at lower levels in the hierarchy of replicators because the rates of turnover for lower units are greater, and heritability of these traits is higher.

One of the most common misconceptions about natural selection is that the process operates directly for the long-term benefit of a species, rather than for the short-term reproductive interests of individuals. We often hear in popular accounts that males and females come in approximately equal numbers because this minimizes the general level of strife within a species over mate acquisition; that organisms voluntarily withhold reproduction at higher population densities to avoid an exhaustion of resources that could lead to species extinction; that individuals in older age classes die to make room for the young; and in general that altruistic behaviors commonly evolve because of the collective good that they confer to the species. Such interpretations are anathema to most evolutionary biologists.[10] Natural selection operates by the differential survival and reproduction of individuals, their close kin, and the genes they carry. Any advantages that accrue to the species are mere byproducts of such individual selection, and do not represent primary adaptations that originated through natural selection among higher units of biological organization.

Nevertheless, overall patterns of organismal diversity can be affected by the differential survival and proliferation of species and higher taxa. For example, dinosaurs went extinct (probably as a consequence of an asteroid impact) 65 million years ago, whereas some mammalian lineages survived. Traits that permit survival during such extinction episodes may be considered adaptive in a sense, but often in happenstancial ways because these traits may have originated as neutral features or features serving other functions. If these traits prove to be advantageous in a new environment, they can open up previously unavailable niches to their possessors. Similarly, some adaptive features forged by natural selection within species might incidentally promote speciation events themselves, leading to a synergistic increase in representation among descendants.[11]

A partial analogy may exist between the action of mutation and recombination in conventional adaptive evolution, and that of speciation and extinction in evolutionary sorting at higher evolutionary echelons. Although room remains for scientific debate, most geneticists believe that *de novo* mutations typically arise at random with respect to fitness; in other words, that there is no directional tendency for mutations to be useful or appropriate.[12] Particular recombinational events during sexual reproduction (though not necessarily the composite rate and pattern of recombination) also are believed to be random with respect to adaptive requirements. Natural selection then becomes the sole sculptor of adaptive change within species, utilizing the raw materials ultimately provided by these and related variation-generating processes. Similarly, when environments are altered dramatically and major extinction and speciation episodes thereby precipitated, selection-honed adaptations within species may or may not be useful under the new circumstances. Dinosaurs and mammals both possessed many fine adaptations to late-Mesozoic conditions, but these features could not have anticipated the novel selective challenges presented by the asteroid holocaust.

Another requisite for evolution by natural selection is that the self-replicating units are packaged in ways that allow at least the possibility for long-term persistence. In multicellular organisms such as ourselves, somatic cells have no real future because they will die with the organism. Furthermore, barring the occasional mutation, all somatic cells within an individual normally are alike in genotype. Thus, any differential proliferations of alternative cell types during an individual's lifetime are considered to lie within the realm of ontogeny (development) rather than evolution. This conventional understanding of ontogeny becomes strained, however, when different genetic elements proliferate unequally within the germ-cell lineages of individuals, in which case the genetic changes can be considered microevolutionary.

Natural selection has been described as a "blind watchmaker."[13] It is also an unconscious and amoral watchmaker, totally devoid of intelligence, foresight, and ethics. From among the genetic variants that arise through the hereditary processes of mutation and recombination, natural selection in effect makes choices based solely on present reproductive performance, without regard to future prospects, potential for subsequent improvement, or any explicit aspects of individual well-being (except insofar as they contribute to fitness). Yet from out of this mindless process all forms of life have emerged, including a species *Homo sapiens* with the very characteristics that natural selection lacks: intelligence, vision, and the capacity for perceptions of moral conduct. This revelation is disturbing to those who understandably gain far greater comfort from believing in an omniscient power. However, the evolution of humankind through natural causes can be viewed with a sense of awe and inspiration too, perhaps even more so than had we merely been created under the direct auspices of a deity. Many biologists and others may agree with the concluding sentiment in Darwin's treatise: "There is grandeur in this view of life, with its several powers, having been originally breathed by the Creator into a few forms or into one . . . from so simple a beginning endless forms most beautiful and most wonderful have been, and are being evolved."[14]

The third doctrine of twentieth-century biology is historicity, the contingency of evolutionary outcomes on prior events. The streams of heredity that connect all present-day life forms to their ancestors have meandered through only a small subset of the conceivable mutational and selective pathways.[15] This dependency on unique genetic and environmental events implies that evolved organismal features are inherently unpredictable in detail. Organisms can be expected to display adaptations to their environments, but the niceties and minutiae of particular adaptive features could not have been (and cannot be) predicted with any great accuracy. Furthermore, upon close inspection, many organismal adaptations appear jury-rigged rather than intelligently designed. The historicity of evolution dictates that natural selection fashions outcomes only from preexisting biological fabrics that in turn had idiosyncratic historical antecedents.

Like other organisms, humans have their share of phylogenetic legacies that constrain adaptations far short of designer perfection. An excellent example of such a design flaw concerns an unwanted junction of our food-conveying esophagus with our air-conveying trachea (thus posing ever-present dangers of choking on food). How insensible it is that these two thoroughfares should intersect, necessitating constant attention by a highly conscientious but nonetheless fallible crossing guard (the glottis). A more intelligent solution to this engineering problem would entail complete separation of the respiratory and digestive systems, which is the case for insects and mollusks. It is something of an accident of history that the early ancestor of vertebrates was a small aquatic creature whose oral cavity simultaneously served as a feeding sieve and a gill apparatus for extracting oxygen from water. Hundreds of millions of years later, our respiratory and digestive systems still retain this legacy.

Another well-known maladaptive legacy is the human appendix, which has no known positive function, but certainly can bring agony and quick death upon rupture. This troublesome outpocket of the large intestine is the vestige of a digestive organ, the caecum, that in other mammals such as rabbits serves to process low-nutritional plant substances that became less prominent in the diets of our primate ancestors. If the appendix now confers no benefit to humans, but has obvious fitness costs, why hasn't natural selection resulted in its complete elimination? Evolutionary biologists recently raised an intriguing possibility.[16] Perhaps natural selection can reduce the size of the human appendix only to a point. A further narrowing and constriction of the appendix might in-

crease the risk of fecal infection and appendicitis, thereby paradoxi-
cally selecting for maintenance of an intermediate but still less-
than-useless vestigial trait.

There are numerous other historical design flaws in the physiological and mechanical makeup of human beings, such as a metabolic inability to manufacture vitamin C; the absence of a reserve
second heart (unlike our paired lungs, kidneys, eyes, and opposable
thumbs); a birth canal too narrow to permit comfortable passage
of an infant; the retention of wisdom teeth in a jaw that is too
short; and the prevalence of problems that accompany upright
bipedalism, ranging from pains in the lower back, leg joints, and
feet, to abdominal hernias, varicose veins, and hemorrhoids.[17] Such
imperfections of design, all too familiar especially to those of us of
advancing age, are as understandable in the light of evolutionary
history as they are unfathomable as the workings of a loving
interventionist god.

The historicity inherent in the evolutionary process implies that
if the tape of life on earth somehow were to be replayed, the movie
would have different actors and plot.[18] Yet the script would pro-
ceed under the sponsorship of genetic variation, the production
services of hereditary mechanisms, and the directorship of natural
selection. There is an inevitability to adaptive evolution, but there
is no inevitability to particular evolutionary outcomes, including
the appearance of intelligence or of human life. From this perspective, we should cherish human existence all the more.

Evolutionary Genetics and the Human Experience

The genetic gods have arisen and been modified through evolutionary processes. Any comprehensive attempt to understand the
nature of these gods (and hence of ourselves) requires a strong
evolutionary orientation. This is a relatively new thesis in the
history of philosophy and thought about human nature, a thesis
that can be and has been much abused in the century since Darwin.

Early in this century "social Darwinists" rationalized excesses of
the free enterprise system (such as sweatshops for children) as a
natural consequence of survival of the fittest in an economic arena.

Eugenicists of the 1930s believed that some human races or social groups were genetically superior to others, with such horrific social consequences as the Holocaust. Many abuses of evolutionary genetics are perpetrated by those with ulterior social or political motives who are poorly informed or misinformed about the evolutionary-genetic processes they appropriate. Recognition of the obvious potential for abuse of evolutionary-genetic principles has made the public, and even some scientists, so wary that evolutionary research often has been stalled or halted altogether.[19]

In addition to concerns about the potential for political or social abuse, there are other reasons why an evolutionary perspective on the human condition must be approached with caution. In most cases, any genetic influences underlying particular human behavioral states are still rather poorly understood. Although reports of newly identified genes for certain behaviors appear almost weekly in scientific journals, current understanding of evolutionary principles and processes derives primarily from comparative studies of nonhuman animals and plants.

Few topics run a greater risk of generating pointless contention than evolutionary genetics. Some of these issues I've chosen not to tackle in this book. I do not, for example, propose that a scientific understanding of human origins and nature will necessarily improve the human condition, either with regard to the collective good of the species or the private fulfillment of the individual. The primary goal of basic (as opposed to applied) science is to pursue paths of objective reality regardless of where they may lead. In this respect, it is unfair to judge the immediate products of pure scientific inquiry against those of religious or philanthropic pursuits. The insights of basic science frequently can be and have been used to alleviate human suffering or to better understand the human condition, but the scientific exercise of understanding the nature of life must be set apart from practical concerns.

Although I speculate on the basis of evolutionary biology about how the human ethical perceptions and behaviors that we exhibit may have come into being, I make no claims that what is evolutionarily natural is in any way morally right. Science is descriptive rather than prescriptive, and I leave to others the task of considering whether an evolutionary understanding of human nature might be used to better the human condition.

In this book I speak of "genetic determinism," but not in the pejorative sense often associated with the nature/nurture debate. The sharp divisions drawn in this debate too often miss the point that all organismal features result from an interaction between genetic makeup and environment. Every suite of genes requires some suitable range of environmental conditions for proper expression, and every environment will suit some genotypes better than others. When things go awry and genetic disorders appear, these may be interpreted as an incongruence or an improper match, between genes and the environment. True, certain traits tend to appear across a wider range of environmental conditions than others. Yet, for any genotype, some environments will be entirely inhospitable.

Furthermore, even traits whose expression is influenced by circumstance have a genetic basis. For example, the particular language spoken by a human child is directed by societal context, but the capacity to acquire elaborate language skills remains a genetically-based feature specific to humans. Many human characteristics are likely to have such deeper genetic underpinnings notwithstanding variable expression as a function of societal context. In general, considering gene-environment interactions in trait expression provides a preferred framework for discussions of human nature. Thus, biological determinism includes both direct influences of genes on human nature and indirect influences of genes through environmental and cultural settings and thought processes that the biologically-produced human mind enables. Under this broader definition, biological determinism might be contrasted more appropriately with metaphysical than with cultural determinism.

The existence of evolution is not under debate by most biologists, but evolution's ramifications for human affairs are largely unexplored. This book is intended to stimulate exploration of that field and to facilitate communication with theological inquiry rather than to evangelize particular scientific canons or unduly demean nonscientific ones. This book is not a call to atheism, either in spirit or practice. I hold no illusions that even a prolonged exposure to science would convert most people who hold traditional religious beliefs. In 1916, James Leuba conducted a landmark survey of a thousand practicing scientists on matters of theism, and found that about 40 percent of those questioned retained belief in a personal god and an afterlife. Leuba predicted that such faith would erode dramatically in this century as scientific understanding grew, but a comparable survey conducted eighty years later showed that a nearly identical percentage of current scientists still believe in a god to whom one may pray in expectation of receiving an answer.[20] Clearly, Leuba had misjudged either the ability of science to satisfy all human needs, or the power of the theistic hold on the human mind.

Although many religious beliefs are flatly incompatible with open-minded scientific inquiry, religion and science need not be at odds.[21] Evolutionary and genetic findings could provide the religious with rich material for enlightened deliberations on the human experience, and biologists might take a more active interest in exploring with nonscientists the possible ramifications of genetics and evolution for the human condition. In a recent statement, Pope John Paul II suggested such a possibility: "It is crucial that this common search based on critical openness and interchange should not only continue but also grow and deepen in its quality and scope. For the impact [science and religion have] and will continue to have, on the course of civilization and on the world itself, cannot be overestimated, and there is so much that each can offer the other."[22]

TWO

Geneses

A myth is a projection of an aspect of a culture's soul . . . [It] is
to a culture what a dream is to an individual.

D. A. Leeming and M. A. Leeming,
Encyclopedia of Creation Myths

In the last two hundred years, science has shed much light on
several geneses that have profound relevance for theism (or belief
in anything supernatural): the origin of replicating molecules
from a primordial soup some three to four billion years ago; the
gradual evolution of human genes from those of primate ancestors
beginning some five million years ago; and the gene-programmed
development of the individual from a fertilized egg. If a god is
monitoring these scholastic achievements, and if this god values
open-minded inquiry and knowledge, then the almighty god must
be pleased with this unprecedented scientific progress by human
disciples.

Humans always have been fascinated if not preoccupied by
questions of our place in some broader scheme. Nearly all major
and minor religions have creation stories: of individual and tribal
origins, human origins, or the origins of life and the universe.[1]
Many of these stories are poetic, beautiful in spirit, and emotionally
evocative. Typically, they are meant to make an astonishing world
less strange, or to bring meaning, order, and certainty to what
otherwise can appear frightful and chaotic. Yet they remain myths.
Only in the past century and a half have scientific modes of inquiry
been adopted that move from traditions of imaginative storytelling,
soothsaying, and religious revelation toward objective observation,
experimentation, and critical evaluation. Many of the scientific

discoveries concerning biological and human geneses have come from the fields of genetics and molecular evolution. These findings too can be viewed as beautiful, poetic, perhaps even uplifting to the human spirit, although such subjective criteria cannot be a basis for their evaluation as truth. The goals of this chapter are to provide examples of traditional myths about biotic origins, to highlight some of the pertinent findings on these origins from evolutionary genetics, and to consider philosophical responses to the new scientific discoveries regarding our perceived position within the natural world.

Mythologies of Biotic Geneses

There are hundreds of creation stories which have been told and retold throughout recorded times by various human cultures. No doubt most familiar to Judeo-Christian readers are the biblical accounts of creation as presented in Genesis. Here, in the first few days God created the heavens and the earth from darkness and void. On days three and five, respectively, God populated the world with plants and then animals. On day six, God created the first man (Adam) and woman (Eve), the latter from one of Adam's ribs. This account is one of many such examples of creation ex nihilo—in this case via the edicts of an omnipotent deity.[2] The account also entails an implicit concept of time as an arrow (unlike time's cyclical nature in Buddhism).

Actually, the biblical account contains two distinct narratives: a story of the origin of the world and its life (Genesis 1:1–2:4) as compiled by a succession of priestly scholars beginning in the sixth century B.C.E.; and the story of humanity's creation and early pedigree (Genesis 2:4–3:24) as had been related by a poet-storyteller approximately four centuries earlier. At the time of the scholars' writings, the monotheistic Israelite nation faced exile in Babylon. To preserve and distinguish the Israelite tradition from the capricious and magic-filled scenarios of other Near Eastern creation myths of the time, these scholars portrayed a single god, almighty and untouchable, who dictated a grand and structured origin for a perfect human species, fully blessed and made in his likeness. By contrast, the poet-storyteller version in Genesis tells of a not-so-perfect origin of humans from dust, dictated by an anthropomorphised god who tempted and then punished his creations in an evocative drama complete with a talking serpent, a paradisiacal garden, and developing tensions between man and woman, humans and the environment, and most important, between humanity and God.

These biblical scenarios illustrate the kinds of contextual orientations that characterize many creation myths. For example, the desert environments inhabited by the Canaanites of Babylonian times may have prompted the Genesis metaphor of the Garden of Eden as a fruitful oasis, humanity's womb. More generally, the sheep-herding lifestyles and traditions of the Middle East probably fostered a worldview reflected in one of God's edicts to man (Genesis 1:28): "fill the earth and subdue it; have dominion over the fish of the sea, over the birds of the air, and over every living thing that moves on the earth." Today, the word "pastoral" refers equally to a shepherd's way of life, or to the relationship of a pastor (shepherd) to his congregation (flock).

Environmental and societal contexts play an even more evident role in many other creation myths. The Micronesians of the Marshall Islands tell of an ancient time when there was only water, and the god Lowa was alone. Lowa hummed, and islands, reefs, and sandbanks emerged. He hummed again and plants and animals appeared. Lowa then created a man, who placed the islands into a coconut-leaf basket and spatially arranged them into the present-day Caroline and Marshall archipelagos. Lowa then sent tattooers to the islands to give all of their various creatures distinctive marks.

The whale hunters of Kukulik Island in the Bering Sea tell of a Creator-Raven who first made land and shores, and then reached down into deep water for some pebbles from which he made people whom he taught to pick seaweed, hunt, and fish.[3] One day a man asked the Sun for some reindeer, but was given special pebbles instead. When these were thrown into the sea, they became whales.

In the creation story of the Papuan Keraki tribe of New Guinea, the first humans came out of a palm tree when the god Gainji heard them speaking in many languages from within. After they were freed from the palm trunk, the people of different language groups went their separate ways, and so it is today. For the Sumu peoples of Central America, life began when two deistic brothers created a beautiful physical world. When admiring their work by canoe, they fell into a rapids, and swam shivering to shore where they built a fire for warmth. One of the brothers (Papan) then found some cobs of maize, which he threw to the ground to produce animals, and into the water and air to produce fish and birds. Enraptured by this diversity of life, the brothers then stumbled back into the fire and were consumed by flames. The sparks from one brother became the stars, whereas those from Papan became the sun (or Sun-Papa) from which all Sumu people have descended.

It is entirely understandable that human cultures should sculpture creation myths from the landforms and biotas familiar and important to them. Social environments too have influenced (and in turn been influenced by) the types of myths formulated. One example involves the Hopi Indians of the American Southwest, whose society is strongly matrilineal (property and individuals belong primarily to the mother's family). Correspondingly, Hopi myths of origin are dominated by the actions of a female creative principle Hurúing Wuhti—variously referred to as "Hard Beings Woman," "Earth Woman," or "Spider Woman"—who by sacred thought originally formed a man and woman and cradled them in her arms until they breathed life.

An elaborate and imaginative creation story was told by a shaman of the Thompson Indians of British Columbia. In the beginning, only water existed, and so a bored god ("Old One") plucked five hairs from his body that transformed into five beautiful and spirited young women. The first of these women declared a desire to pursue pleasure, and to have many wicked children who would become fighters, adulterers, liars, and thieves. The second also wished to bear children, but these would be wise, honest, chaste, and peaceful. The third young woman wanted to be a nurturer who would give abundantly of herself; the fourth wished to generate feelings of warmth; and the fifth simply desired grace and fluidity. Old One granted all of their wishes. The fifth woman became water; the fourth became the spirit of fire in all things; and the third young woman became the earth mother. This explains why we are all directly related to water, fire, and earth. The first and second women were impregnated by Old One, and sent to populate the world. Old One foresaw that children of the evil woman would predominate at first, but eventually the good woman's descendants would prevail. This explains why there are both good and bad people. Someday, at the end of the world, Old One will bring together the five women and all good and evil people, both living and dead.

The creation myths of a few cultures have evolutionary elements. For example, some Indian tribes in southern California firmly believed that humans were descended from animals. According to one account, the ancestors of mankind were coyotes who developed an odd habit of burying their dead. This custom induced changes in the coyotes themselves: By sitting upright at burial ceremonies, the coyotes eventually wore away their bushy tails. Many generations later, after they began standing, forepaws lengthened into human hands and muzzles gradually shortened into human faces.[4]

Several recurring themes emerge from a comparison of creation myths across cultures: a common assumption that creation began near the center of a culture's local world; as mentioned, the incorporation of familiar environmental conditions (e.g., creatures, landforms, societal mores) into the creation stories; a concept of time as a linear thread for life starting with creation and often projected to end with some sort of final reconciliation or judgment; and the involvement of one or more powerful deities. The creations are of something from nothing (material from the ethereal), order from chaos, of life from nonlife, of a particular tribe, or of the entire human race. The creator(s) may be the allegorical likenesses of human father figures or mother figures, various animals such as the raven or turtle, inanimate entities such as water, thunder, and the sun, or cosmic eggs (precreation voids that in many myths bear some analogy to the observable eggs of birds and reptiles). The creators may be all-powerful or mortal, honest or tricky, altruistic or selfish. The creations may be achieved by mere thought or will, verbal commandment, physical manipulation of materials such as clay or dust, or by deistic dismemberment, secretion, or defecation.[5]

Few of these creation stories are flatly incompatible with current scientific knowledge provided they are to be interpreted in an allegorical sense. For example, in the light of scientific evidence, the whale pebbles of the Kukulik Islanders might be reinterpreted as a return to the sea of terrestrial bodies ancestral to cetaceans; or, the seven days of creation in Genesis might be viewed metaphorically as evolutionary stages in the appearance of nonhuman and then human life. However, conflicts between faith and science can arise when some of the traditional creation myths are taken too literally. Against current scientific understanding, it is no more plausible that pebbles thrown into the ocean spontaneously produce whales than it is that the genesis of humans took place in a literal twenty-four–hour period beginning at 9 A.M. on October 23, 4004 B.C. (a date calculated from biblical accounts by the seventeenth-century divine, Bishop Usher).

More reasonable spiritualists have less problem accommodating creation myths, viewed allegorically, with scientific advances. For example, most practicing Christians today have no particular difficulty in harmonizing their spiritual views with the scientific facts that the earth is neither flat nor the center of the universe. Even so, such conciliation sometimes comes slowly or begrudgingly. In a famous inquisition by the Catholic Church in 1633, the scientist and scholar Galileo was found guilty of heresy for his discovery that the earth orbits the sun, and hence was not the center of creation. Galileo, then an old and ailing man of sixty-nine, was ordered upon pain of death to recant his views. When he died a few years later, the Church refused to allow his burial on consecrated ground. It was not until 1993, 360 years later, that a Vatican headed by Pope John Paul II officially admitted the Church's long-standing error. For some, the facts of evolution are even more bitter pills to swallow.[6]

Science of Biotic Geneses

Against this backdrop of the millennia-old preoccupation with creation myths, the level of understanding of biotic geneses achieved within the last 150 years of biological science truly is remarkable. Prior to about 1850, many members of both literate and preliterate cultures believed that contemporary organisms originate by spontaneous generation from nonlife (e.g., frogs and turtles arise from mud, and birds now understood to be migratory emanate from water or forest detritus each spring), that humans bear no genealogical relationship to other animals, and that conception and pregnancy sometimes are mediated by spirits. The contrast between the texts and stories of folklore on the one hand, and current scientific treatises on biology, genetics, and evolution on the other, could hardly be more striking.

Genesis of Primordial Genes

The earth, formed by the gravitational accumulation of matter from the disk-shaped swirling cloud of our primordial solar system, is approximately 4.5 billion years old.[7] It is one of nine known planets orbiting the sun, which is one of billions of stars in the Milky Way, one of millions of galaxies in the known universe. On this tiny speck of matter, forms of life appeared some 3,500 million years ago. Evidence for these early microbes consists of cellularly preserved prokaryotic microfossils and stromatolites[8] recovered from ancient sedimentary deposits in South Africa and Australia. It remains to be discovered whether life exists elsewhere in the universe, and, thus, whether biotic geneses are rare accidents, or virtual certainties wherever suitable prebiotic conditions exist. In any event, within a "mere" 1,000 million years of the solar system's origin, abundant microbial life on this planet was off and swimming.[9]

How this life arose from nonlife is a scientific mystery about which there is considerable educated speculation and growing experimental evidence. Typically, four stages in the process are envisioned: the accumulation of small organic building blocks including amino acids and nucleotides; assembly of these units into proteins and nucleic acids; aggregation of these polymers into "protobiont" molecules; and the origin of hereditary mechanisms. In the last fifty years, elements of all four stages in the dawn of the primordial genes have been reproduced experimentally in the laboratory.

Following suggestions originally made in the 1920s, Stanley Miller in 1953 reported the first laboratory experiments to spontaneously generate organic molecules from inorganic compounds. Miller attempted to mimic environmental conditions thought to have been characteristic of the primordial earth.[10] Into sealed flasks containing water (a miniature sea) and hydrogen, methane, and ammonia (a primitive atmosphere), sparks were discharged to simulate lightning (see Figure 2.1). After one week the circulating, heated solution was found to contain a variety of organic compounds, including the amino acids alanine, glycine, aspartic acid, and glutamic acid. Many variations on this basic experimental design have been conducted since, and from such chemical alchemy has emerged a veritable litany of organic compounds: di- and tri-carboxylic acids, branched- and straight-chain fatty acids, fatty alcohols, sugars, triazines, imidazoles, and a host of others.[11] Most germane to the issue of life's origin on a primeval earth, these

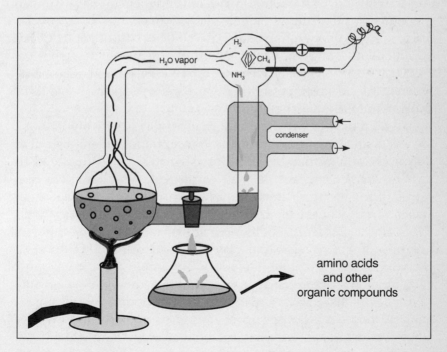

Figure 2.1 Laboratory apparatus used by Stanley Miller in 1953 to simulate conditions of the Earth's early atmosphere. These experiments were the first to demonstrate the abiotic synthesis of organic molecules.

experiments also have generated the twenty amino acids that are the most common components of proteins today, as well as the purine and pyrimidine bases in the nucleotides of modern nucleic acids. In the design of a typical experiment, free oxygen and (of course) microbial life are excluded because these would have been absent in the primitive earth, and also because they would tend to oxidize and eat, respectively, the developing organic soups.[12]

In living organisms, enzymes catalyze the assembly of proteins and nucleic acids from monomeric subunits. However, abiotic polymerizations may have taken place on a primitive earth, particularly if the brewing organic soup was sufficiently rich or chunky to regularly bring squadrons of amino acids or nucleotides into close proximity. Spontaneous abiotic polymerization of "proteinoids" has been demonstrated in the laboratory when organic solutions are dripped onto hot sand or rock (vaporization of water concentrates the monomers). Natural analogues might have been monomer-laden water droplets splashed from hot springs onto adjoining rocks, or from oceans into sun-baked tidepools. Other proposed concentration vesicles include proteinoid microspheres, colloidal droplets called coacervates, and even lipid membrane spheres. All of these possibilities have received some degree of experimental laboratory support.

In one of the more intriguing scenarios, clay plays a pivotal role. Clay is a suitable substrate for the concentration and subsequent polymerization of organic monomers because the latter tend to bind to charged sites on the clay particles. In traditional creation myths of many human cultures, life's origin from clay is also an ancient and recurring theme. Clay is a preferred substance from which the deities construct living matter. Wouldn't it be wonderfully ironic if clay really did play a pivotal role in life's natural creation?

Another probable key to life's origin is the emergence of organic molecules with catalytic capacities. In the modern world, enzymatic proteins coded by specifiable genes are by far the most widely employed of organic catalysts, facilitating everything from DNA replication to the digestion of candy bars. Even simple polypeptides produced abiotically sometimes display weak catalytic properties, suggesting to many researchers that proteins were probably the first living molecules. An alternative school of thought champions polynucleotides (or, more precisely, poly*ribo*nucleotides, the building blocks of RNA) as the first molecules of life.[13] In favor of this view is the recent revolutionary discovery that some RNA segments (known as ribozymes) themselves have an intrinsic capacity for efficient biological catalysis, including the autocatalytic function of directing further RNA synthesis.[14] To many scientists, the concept of an early "RNA world" is appealing because it might help to solve the chicken-and-egg dilemma of how an emerging life form could self-replicate in the absence of preexisting catalytic machinery. Yet the "protein-first" and "RNA-first" hypotheses are not mutually exclusive. Stuart Kauffman suggests that life truly blossomed when ubiquitous polypeptides *and* polyribonucleotides with catalytic capacities became coupled into complex polymers with both coding and self-replicating functions.[15]

After variable organic molecules with self-replicating capabilities were on the scene, natural selection could truly grab hold to promote biotic adaptations to the physical environment. As life diversified, so too did selection pressures, affording yet more challenges and opportunities for biotic proliferation in a positive feedback process that still fosters biodiversity today. By about 1,400 million years ago, microbes had assembled into the first eukaryotic cells with differentiated organellar functions. About 700 million years ago came the first multicellular animals, mostly flat and soft-bodied, and within another 150 million years the first hard-bodied creatures had made their appearance in substantial numbers and variety.

Although many of the organic components of life have been demonstrated experimentally to arise from inorganic materials under suitable laboratory conditions, *de novo* geneses of complex genes displaying the full-fledged capacity to direct and replicate life have not yet been synthesized by human hands. Some might argue that this disproves a natural origin for life, but such conclusions are unjustified considering the temporal and spatial scales available for nature's experimentations. Human efforts at the experimental synthesis of life have been conducted over fifty years, whereas nature had 1,000 million years to work with, a 20-million-fold difference. Generously, human efforts may have involved a hundred square meters worth of experimental flasks, whereas nature had available the 500,000 million square meters of the earth's outer surface, a 50-trillion-fold difference. Thus, the temporal and spatial opportunities for life's genesis in the primeval earth were astronomically greater than those available under recent human auspices.[16] From this perspective, perhaps most surprising is not that full-blown life has yet to be synthesized artificially from inorganic compounds in the laboratory, but that so much progress toward that end has been achieved by scientific experimentation in such a short order.

Genesis of Human Genes

The human species is one of about 1.5 million described species currently inhabiting the earth. In most traditional classifications, *Homo sapiens* is the sole living species assigned to the Hominidae, a taxonomic family in the superfamily Hominoidea that also includes the Asiatic apes (gibbons, siamangs, and orangutans) and the African apes (chimpanzees and gorillas). At face value, this classification implies that although humans have evident similarities to the great apes, our biological differences nonetheless are pronounced enough to place *Homo sapiens* into a monotypic family (a rather unusual honor in the broader practice of animal systematics). Unfortunately, the fossil record for humans and the great apes is notoriously poor, and gaps remain in our understanding of human origins from fossil evidence alone.[17]

Scientists have filled many of these gaps by developing laboratory technologies that permit assays of genes and their protein products.[18] In the last thirty years, molecular geneticists have learned how to read the genetic scriptures of humans and other species directly. Not only do these DNA scrolls provide the coded prescriptions for life, but they also contain detailed sagas of the evolutionary histories of genes. A close reading of the genomic scriptures of humans and other primates has revolutionized understanding of the recent genealogical past of *Homo sapiens*.

Those unfamiliar with the workings of a modern genetics research laboratory often envision scientists as peering endlessly through microscopes to decipher the characteristics of DNAs or proteins. Actually, the relevant textual features of these biological macromolecules are far too small to be seen directly under even the most powerful of microscopes. The trick thus has been to scan the genetic scriptures through biochemical means, and to project the results onto electrophoretic gels or autoradiographs that display the genetic texts in a format that can be read by the human eye or computer scanners.

One of the earliest and simplest of the biochemical techniques to scan genetic information, protein electrophoresis, was deployed widely in the mid 1960s. Small samples from the heart, liver, or other tissues are homogenized in a liquid and centrifuged to pellet the unwanted cellular debris. A droplet from the top layer, which contains the water-soluble proteins of interest, is then applied to a paper wick and inserted into a gel often made from potato starch or acrylamide. The gels are much like Jello: ingredients are boiled, poured into a mold (in this case, a centimeter-thick slab the size of stationery paper), and allowed to set. Next, an electric current is applied that causes the thousands of proteins in the original droplet to migrate through the gel at rates determined by their molecular configuration and net charge, which vary from one protein to another. Biochemical stains then are applied to illuminate particular protein bands in the gel. Genetic inferences are made by comparing the band profiles from one individual or species against another.

Many other molecular assays are variations on this theme. Biological macromolecules are isolated from tissue samples, electrophoretically separated through gels, visualized as bands, and genetically interpreted.[19] In one of the more widely used modern assays, the gel bands represent ordered sequences of nucleotides in particular genes. A DNA sequencing gel can reveal several hundred nucleotide characters at once, the genetic equivalent of about four or five lines of this book. By such molecular transliterations in the laboratory, entire sentences, paragraphs, and chapters of the genetic scriptures gradually come into view.

A stunning and incontrovertible outcome from three decades of this molecular analysis is the close similarity of human genes to those of the great apes, notably the gorilla *(Gorilla gorilla)* and chimpanzees *(Pan troglodytes* and *Pan paniscus)*. Protein electrophoretic assays have shown that humans and chimpanzees are about as similar to one another genetically as are morphologically similar species of fruit flies (genus *Drosophila*), or sunfish *(Lepomis)*.[20] From a fly's or a fish's perspective, humans and chimpanzees therefore might warrant placement in a single genus, *"Homopan."* Gorillas too have proved roughly equidistant genetically to chimpanzees and humans in these assays, such that a genus *"Homogorillapan"* might be contemplated. The close similarity between humans and the great apes is not an artifact of the limited resolution of protein electrophoresis. Direct assays have shown that the amino acid sequences of typical human and chimpanzee proteins are more than 99 percent identical.[21]

These conclusions about the molecular similarity of humans to the great apes have gained further support from results of DNA-level assays of genes themselves.[22] The most extensive nucleotide sequence information has come from the rapidly evolving mitochondrial genome, where complete DNA sequences have been obtained from humans, chimpanzees, the gorilla, and other great apes.[23] Among the more than 16,000 nucleotide positions in this molecule, only about 1,400 (8.8 percent) differ between humans and chimpanzees. Comparable sequence divergence estimates between humans and gorillas, and between gorillas and chimpanzees, are 10.6 percent and 10.3 percent, respectively. Within this triad of species, humans and chimpanzees appear most similar. The evolutionary rate of mitochondrial (mt) DNA in mammals is known to be about 2 percent sequence divergence between a pair of lineages per million years.[24] Under this evolutionary clock, the phylogenetic separation between humans and chimpanzees is estimated to have taken place about 4.5 million years ago.

Apart from such direct sequence comparisons, available only for small numbers of hominoid genes, a "DNA–DNA hybridization" approach permits a quantitative appraisal of composite sequence divergence across the entire nuclear genome. In this interesting procedure, "single-copy" DNAs isolated from species A and B are denatured into single strands that then are allowed to reassociate in a test tube into "heteroduplex" DNA molecules with strands from species A matched to counterpart strands from species B. In temperature treatments, the thermal stabilities of these heteroduplexes are compared to those of control homoduplex DNA molecules, where both paired DNA strands are from the same species. Heteroduplex DNA molecules typically display lower chemical stability at elevated temperatures than do homoduplex DNAs because fewer of the nucleotides align properly and, thus, fewer chemical bonds bind the two strands. Higher thermal stability indicates that the species compared are more alike. For the human-chimpanzee heteroduplex molecules, the magnitude of depression in thermal stability indicates about 1.6 percent nucleotide sequence difference between the DNA strands from the two species. Under appropriate molecular clock calibrations, this value translates into an estimated evolutionary date of about 5.9 million years ago for the phylogenetic split leading to humans and chimpanzees.[25]

As creationists routinely point out, the close genetic similarity between humans and the great apes is not alone definitive evidence of shared evolutionary ancestry. Under their interpretation, close resemblance merely reflects a creator's choice to produce similar organisms in separate creations. This explanation poses a philosophical quandary for special creationists, as George Romanes, a scientific defender of Darwin, pointed out more than a hundred years ago: "If we reject the natural explanation of hereditary descent from a common ancestry, we can only suppose that the Deity, in creating man, took the most scrupulous pains to make him in the image of the ape . . . Why should God have thus conditioned man as an elaborate copy of the ape, when we know from the rest of creation how endless are His resources in the invention of types?"[26] For the sake of argument, let us assume that God had reasons for making humans and apes similar. Plausibly, these independent creative endeavors might have required similar molecular structures. There is a logically compelling counter-argument to this proposition, based on a precedent established in the legal profession.[27]

Publishers frequently bring suit against suspected plagiarists, particularly when popular textbooks such as those in elementary chemistry or biology are involved. A routine defense is that these subject matters are circumscribed and standard, such that similarity in textual treatment results from "separate creations" by the authors rather than plagiaristic copying from "ancestral" texts. Courts generally have upheld this interpretation, but there is one important exception. When particular *mistakes of detail* in an original text are faithfully repeated in later treatments by separate authors, copyright laws deem it inconceivable that identical errors were made independently by the plaintiff and defendant. (Indeed, publishers sometimes set up sting operations for plagiarists by including a few false entries.) It's not hard to see the similarity of this situation with the special-creationist versus evolutionary scenario for molecular mistakes in the genetic scriptures. Organismal genomes are riddled with functional "errors." Yet, fine details of molecular error commonly recur in similar species, effectively eliminating special-creation explanations in favor of historical ones.

One of the more egregious classes of such molecular erratum involves functionless pseudogenes (discussed further in Chapter 4). These gene corpses often are shared by and display fine details of similarity across species suspected to be related. For example, both humans and great apes possess a pseudogene related to a functional gene encoding immunoglobulin epsilon (an antibody protein involved in allergic responses). Furthermore, the pseudogene appears in precisely the same location in the genomes of humans and chimpanzees, an exceedingly improbable outcome under a hypothesis of independent origins. Another example involves pseudogenes in the nucleus apparently transferred from the mitochondrion. Today, these "fossilized," nonoperational nuclear copies of mitochondrial genes are carried by several primates, including humans.[28]

In the light of evolution, such situations are understood simply as a consequence of shared ancestry. In the phylogenetic history of the primates, functionless pseudogenes arose and were copied genetically to descendants. In the light of special creation, the logical explanation must be that a bumbling god repeated precisely the same detailed mistakes multiple times in independently constructed species. In this and many similar cases, an impartial judge in a copyright courtroom could only decide against the special-creationist interpretation and in favor of the evolutionary–genetic view.

The fact that genetic lineages leading to humans and chimpanzees separated from a common ancestor some five million years ago does not imply that modern humans (or chimpanzees) suddenly emerged as we know them today. Rather, evolutionary changes accumulated through a series of transitional populations, which can be traced through hominid fossils. The resulting picture of proto-human structural morphology proceeds from the early Australopithecines to later named species of *Homo* (such as *habilis* and *erectus*) and eventually to the subspecies *Homo sapiens neanderthalensis* and *Homo sapiens sapiens*.[29] Molecular studies of genes have contributed much to the scientific understanding of human origins. Perhaps the most exciting of recent molecular discoveries is that extant human "races" are almost entirely similar genetically. Some of the first evidence came from protein electrophoresis, wherein Caucasoid, Mongoloid, and Negroid populations proved to share identical allelic forms at most surveyed genes. Calculations based on the small mean genetic distances observed led M. Nei and A. K. Roychoudhury[30] to conclude that human racial separations occurred within approximately the last 25,000–140,000 years. Indeed, the great majority (some 85 percent) of the total protein-genetic variation in the human species is intra- as opposed to interracial. One useful way to interpret this observation is to appreciate that if all human populations were to go extinct except for one, and if that population then repopulated the earth, fully 85 percent of the total protein-genetic diversity in the human species would be recovered. Regardless of ethnic background, humans are remarkably alike genetically under their superficially varied exteriors.[31]

The genealogical closeness among human "races" has been confirmed from detailed molecular studies of mitochondrial DNA. As elaborated in Chapter 3, mtDNA evolves rapidly and is inherited matrilineally, without genetic recombination. This means that mtDNA genotypes record maternal family names, and in a fashion analogous to how surnames register paternal lineages in many human societies.[32] Just as sons and daughters receive a paternal surname that sons alone traditionally pass to their children, so too do offspring of both genders inherit mom's mtDNA, which only daughters transmit to progeny. The surname/mtDNA analogy may be carried further. Occasionally, spelling errors arise in surnames. For example, my own surname "Avise" originated in the last century as a misspelling of "Avis" in one small branch of my broader patrilineal family tree. Mutations occur routinely in mtDNA also, and similarly record the histories of matrilines. However, unlike surnames, which are of recent human invention,[33] mtDNA genotypes register a vastly longer history of evolutionary descent with modification.

Through the study of human mtDNA we have found that the history of genealogical differentiation within our species is remarkably shallow.[34] Hundreds of female "family names" have been distinguished, but all are relatively minor variants on a theme. Most mtDNA lineages appear interspersed among the races, and nearly all sequence divergence estimates are well below 0.5 percent (compared, for example, to the human-chimpanzee difference of nearly 9 percent). Based on the mtDNA clock calibration mentioned above, researchers have concluded that all mtDNA lineages present in extant humans can be traced genealogically to a single female ancestor who lived about 200,000 years ago, probably in Africa.[35] The popular press often has misinterpreted this conclusion. Remember that mtDNA chronicles the history of female lineages only. In any population, human or otherwise, some females by chance (if nothing else) will fail to be survived by daughters, whereas others may produce multiple successful female offspring. Thus, as a consequence of population turnover by organismal reproduction, some evolving mtDNA lineages go extinct each generation, whereas others increase in frequency, at least for a time. This lineage sorting process, shown in Figure 2.2, means that any mtDNA evolutionary tree is perpetually self-pruning. It also ensures that, viewed retrospectively, all maternal lineages in any species inevitably coalesce to a single individual at some prior time (every species has its "Eve"). This does not imply, however, that this common ancestor was the only individual alive at that time. To the contrary, she may well have been imbedded within a much larger population of females (and males), many of whom have made genetic contributions to the nuclear genome of the current population.

A mathematical "coalescent theory,"[36] as applied to the empirical mtDNA data for humans under realistic demographic conditions, indicates that our 200,000-year-old mitochondrial Eve was not alone, but rather merely a member of a much larger population probably numbering in the few thousands or tens of thousands of

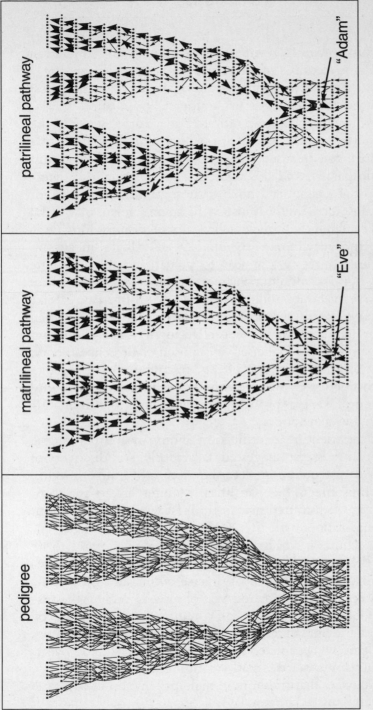

Figure 2.2 Schematic representation of lineage sorting processes and the concept of lineage coalescence to a common ancestor. *Left:* a hypothetical human pedigree or family tree across twenty-two generations, with the current generation at the top. Circles are females, squares are males, and lines connect each individual in each generation to his or her mother and father. *Center:* the same pedigree, but with the matrilineal transmission pathways (female → female → female . . .) highlighted by arrows. Note that all living females trace back through this matrilineal tree to a common ancestor ("Eve") twenty generations ago; note also that Eve was not the only female alive at that time. *Right:* the same pedigree, but with the patrilineal transmission pathways (male → male → male . . .) highlighted by arrows. Note that all living males trace back through this patrilineal tree to a common ancestor ("Adam") nineteen generations ago; note that Adam also was not alone.

individuals. Other lines of evidence also support the conclusion that the total population size of humans never has dropped to a few individuals, even for a single generation. Perhaps the most convincing evidence against a pronounced "population bottle-neck" comes from DRB1, one of about a hundred genes in the human leukocyte antigen (HLA) complex on chromosome 6 that play a molecular role in tissue compatibility and in defense against pathogens and parasites. In this nuclear gene, many alleles (alter-native forms of any gene) are much older than those observed for mtDNA, almost certainly because a balancing form of natural selection has buffered these alleles against extinction for long peri-ods of time. Indeed, among fifty-nine DRB1 allelic lineages in humans, at least thirty-two appear to predate the phylogenetic separation of humans and chimpanzees, meaning that these alleles can be no less than about 5 million years old.[37] If thirty-two DRB1 lineages have persisted for this long, it follows that no fewer than sixteen individuals could have been alive in any generation over that entire span of time. Francisco Ayala used coalescence theory, corrected for probable selection effects, to calculate that the his-torical human population size required to account for the level of DRB1 polymorphism was probably tens of thousands of individu-als on average per generation.[38]

Following the scientific identification of human's mitochondrial Eve, the popular press also suggested, fallaciously, that discovery of our genetic Adam would complete the picture of human origins. The Y chromosome is the patrilineal counterpart to mtDNA. Portions of it are transmitted strictly from fathers to sons (without genetic recombination with the X chromosome). One such Y-chromosome segment, adjacent to the ZFY gene thought to be involved in the maturation of testes or sperm, recently was se-quenced from thirty-eight men of diverse ethnic and geographic origin, and absolutely no variation was observed.[39] Using the sus-pected rate of evolution in the ZFY region, as determined by comparisons of the human sequence against those of the great apes, human's "Adam" was estimated to have lived about 270,000 years ago, which in turn under the mathematical calculations of coales-cent theory suggests that the human male population at this time may have been several thousand.

Although both "Adam" and "Eve" provisionally have been

identified and dated using molecular genetic techniques, it is fallacious to conclude that the picture of human genetic origins is complete. The matrilineal transmission pathway of mtDNA (female →
female → female . . .) and the analogous Y-chromosome patrilineal pathway represent only a minuscule fraction of the total
hereditary pathways available to genes, the vast majority of which
have trickled through the human pedigree through both genders
over the course of multiple generations.[40] Consider, for example,
the origins of your own genes over just the past three generations.
Your mtDNA has come from one great-grandmother, and your Y
chromosome (if you are a male) has come from one great-grand-
father. However, your other genes collectively derive from all eight
of your great-grandparents, having been transmitted under the
rules of Mendelian heredity that ensure an approximately equal
genetic contribution from each of these proximate ancestors. Thus,
before analyses of the genealogical history of the human species
can be considered comprehensive, molecular studies of many more
genes will have to be conducted.[41]

Toward these ends, an academically prompted initiative currently is obtaining and assaying DNA samples from about 10
percent of the world's 5,000 linguistically distinct populations. The
goal of this Human Genome Diversity Project (HGDP, not to be
confused with the Human Genome Project, discussed in Chapter
7), is to provide a globally comprehensive description of human
genetic variation. It is a last-minute effort to archive humans'
genetic history before many local peoples disappear forever or
become genetically assimilated into larger cultures. It should provide a wealth of molecular information about the history of our
species, particularly when integrated with data on language relationships and other cultural characteristics.[42] The HGDP also will
communicate with the Human Genome Project to provide medically important information about the geographic distributions of
genetically-based diseases and human genetic disorders.

Genesis of an Individual's Genes

From numerous observations made through a primitive microscope, the German biologists Matthias Schleiden and Theodor
Schwann concluded in 1839 that all living things consist of cells.[43]

This cell theory, later expanded to include the concept that all cells come from preexisting cells, is a central empirical foundation of developmental biology.[44]

Human development, like that of other multicellular animals, begins when the genes from a female's egg cell are combined with those from a male's sperm.[45] Starting as a single-celled zygote or fertilized egg, each individual then develops by a genetically programmed sequence of molecular and cellular events through the stages of embryo, fetus, infant, child, adolescent, and adult. The process is no less intriguing for its inevitability. To any disbelieving adolescents, it can be pointed out that they will grow old and die just as they grew from an embryo to an adolescent.

Human embryonic development is initiated by successive mitotic divisions of the fertilized egg, a cell about the size of the period at the end of this sentence. First, the zygote divides twice to produce four blastomeres, which then divide again to generate an eight-celled embryo consisting of four sets of cells in two-tiered stacks.[46] Further cell divisions produce a solid ball of cells referred to as a morula (from the Latin for "mulberry"), which soon develops into a blastula consisting of a fluid-filled central cavity (the blastocoel) surrounded by an epithelial cell layer. Next, invagination of one side of the blastula obliterates the blastocoel, producing a cup-shaped gastrula with a new cavity known as the archenteron, destined to become the digestive tract. The advanced gastrula is triploblastic, meaning that one layer of cells (the endoderm) lines the archenteron, another (the ectoderm) forms the embryo's outer wall, and a third (the mesoderm) develops from pouches that bud off the lining of the archenteron. As development proceeds, these primary layers develop into rudimentary tissues and organs. For example, the mesoderm eventually becomes muscle, cartilage, bone, connective tissues, gonads, kidneys, and most of the circulatory system.

From an evolutionary perspective, a particularly important event takes place at an early stage of female embryogenesis when about two thousand amoeba-like cells migrate from the yolk sac to the primordial ovaries of the developing embryo. These cells, which multiply to a few million copies by the time of birth, are destined to become the pool of gametic cells that carry genes of the potentially immortal germ line. However, the odds against transmission to the next generation, let alone immortality, are great: A woman's ovaries are a site of death far more often than of life. Only about four hundred oocyte cells will ripen, to be released one by one during the monthly ovulatory cycle throughout the grown woman's reproductive life. Of these, only a few are likely to be fertilized, and on average, a mere two per female (in a stable-sized population) actually contribute to the gene pool of the next generation. For sperm in males, the odds against success are even greater. A typical ejaculate carries about 600 million sperm, and the cumulative lifetime production by a man may exceed 10 trillion gametes. Yet an average of only about two of these germ-line cells per male will successfully collaborate with egg cells to produce a fetus.

To return to our developing embryo, at five days after fertilization about a hundred cells, all derived from a clonal cellular process known as mitosis, make up the individual. Within four weeks of conception, a human brain begins to take form (from ectoderm), and a tiny heart begins to pulsate. At about thirty-six weeks, an infant is ready to begin an existence outside of its mother's nurturing body.

Nearly all somatic cells of the body contain essentially the same genetic material.[47] Cellular differentiation and organ development therefore result almost exclusively from differential gene expression. Heart and liver cells, for example, express different suites of genes. Recent molecular studies indicate that the number of genes expressed uniquely by brain cells (> 3,000) is far higher than that in any other of the thirty-seven types of human tissue examined.[48] How does such differential gene expression come about, given that all somatic cells of an individual derive mitotically from a single zygote? The picture is starting to emerge as pieces of the developmental puzzle are assembled by biologists working in what has become one of the hottest areas of science. Understandably, most of the research has involved observations and manipulations of nonhuman embryos.

One experimental approach is to transplant nuclei (containing DNA) from differentiated cells into recipient eggs or zygotes from which the nucleus has been removed. In frogs and toads, such transplanted nuclei sometimes direct substantial embryonic development, even to the tadpole stage, but this ability appears to be related inversely to the age of the donor embryos. These experimental results suggest that cell nuclei in these anurans change somehow during development. They also indicate that the changes are reversible to a point, and hence that the differentiated cells probably retain all of the genes necessary for making any frog body part.

A related approach has been to experimentally manipulate entire cells or collections thereof. In the mouse, a blastomere cell isolated from others still can direct the formation of a complete embryo. Such cells are said to be totipotent, meaning that they retain the capacity to differentiate into all other cell types given the proper environment. Human blastomere cells are totipotent as well, at least to the two-cell stage, as evidenced by the routine natural occurrence of monozygotic twins. When mammalian development proceeds beyond the early blastula, conventional wisdom has been that cells gradually lose totipotency as they become more and more specialized. In 1997, however, scientists reported the successful production of a cloned sheep from the nuclear genome of an adult ewe's somatic cell. This astonishing feat indicates that the loss of totipotency through cellular specialization is not invariably irreversible in mammals.

In simplest terms, the genetic riddle of development has been viewed as a problem of deciphering the mechanisms by which cellular specialization is achieved, a problem that fundamentally boils down to one of interactions between the genotype and the environment, the latter writ small. There are many potential sources of microenvironmental heterogeneity (cues) within the embryo to which cellular genomes that are structurally more or less constant might respond during cell differentiation. Egg cells themselves are not mere isotropic and structureless sacs containing DNA. Instead, they represent regionalized chemical information systems that upon division provide different daughter cells with distinctive cytoplasmic environments.[49] In some species such as fruit flies and amphioxus (a primitive chordate), the cytoplasm of the zygote has particularly well-defined poles that set up gradients of intracellular substances called morphogens.[50] These morphogens differentially influence gene expression in daughter blastomeres derived from unequal cell divisions.

After dividing cells begin to adopt different structures and functions, chemical gradients develop in the specialized cellular neighborhoods within the embryo, and these in turn influence subsequent patterns of cell division and tissue morphogenesis.[51] "Induction" refers to the ability of one group of cells to influence the development of another. "Pattern formation," the emergence of specialized tissues and body parts in appropriate locations, must rely also on the proper response by developmental and regulatory genes to "positional information" about a cell's location relative to others. In general, normal cellular specialization in organismal development appears to be a serial process that epitomizes, at the cellular level, the importance of genotype-environment interactions.

Experimental manipulations of embryos have illuminated various aspects of this specialization process at the level of animal morphogenesis. For example, fruit fly larvae contain islands of cells known as imaginal disks that are the cellular precursors of itemized organs in adults. Surgical replacement of a larval imaginal disk specifying "antennae" by an imaginal disk specifying "leg" results in an adult fly with a leg projecting from its head! In other cases, cell bundles involved in animal morphogenesis remain highly sensitive to positional information. If "polarizing activity" cells from a limb bud in a chick embryo are experimentally grafted to a position 180° removed from their usual orientation, the adjacent tissue responds by producing a developing wing that is a mirror image of its normal state (equivalent to a reversal of the palms and the backs of our hands).

At the molecular genetic level, ontogeny is governed by regulatory genes that function throughout the developmental process by controlling the timing of events, making decisions about the fates of groups of embryonic cells, and modulating and integrating the expression of structural genes to produce differentiated tissues. One necessary but surprising feature of development in most multicellular organisms is programmed cell death, or "apoptosis" (from classical Greek, meaning "dropping off"). In fruit flies, for example, a gene known as reaper responds to hormonal and other cellular signals to govern which somatic cells die during normal development. For the greater good of the human individual as well, many somatic cells can and do kill themselves at appropriate times and in appropiate places in the developing body. Such cellular suicides occur routinely, for example, within populations of cells that make up our skin, intestinal and uterine walls, and blood. Growing evidence indicates that many human disorders such as cancers, AIDS, Alzheimer disease, rheumatoid arthritis, retinitis pigmentosa, and osteoporosis sometimes result in part from abnormal regulation or faulty control over the otherwise adaptive apoptosis process.[52]

In general, additional regulatory controls operate at many (nonexclusive) levels, for example during the transcription and translation of genes to proteins (see Figure 2.3), or at any stage after protein production. Informational feedback from the intra- or intercellular environment involves molecular signals that exert direct influence on how genes and their protein products are expressed. Via such molecular avenues, control is exercised over the multitudinous biochemical and developmental pathways to which proteins contribute as enzymatic and structural constituents. The special involvement of regulatory genes in ontogeny also has prompted the now-orthodox view that evolutionary changes in gene regulation probably play a disproportionate role in the emergence of new developmental profiles and the generation of the great diversity of body plans among life forms.[53]

To describe the molecular sophistication and complexity of regulatory genetic control in humans, consider a few of the recently discovered elements in just one aspect of gene regulation: transcriptional modulation (see Figure 2.4).[54] Transcription of a gene to messenger RNA is keyed by an enzyme known as RNA polymerase, which attaches to a core promoter, a sequence adjacent to the relevant structural gene that can be thought of as the ignition switch for a car's engine. Other regulatory sequences called enhancers and silencers, sometimes located thousands of nucleotides upstream or downstream from the core promoter, act as transcriptional accelerators and brakes. Each gene may have several enhancers and silencers, and these can be shared among genes, but different genes have different combinations of them.

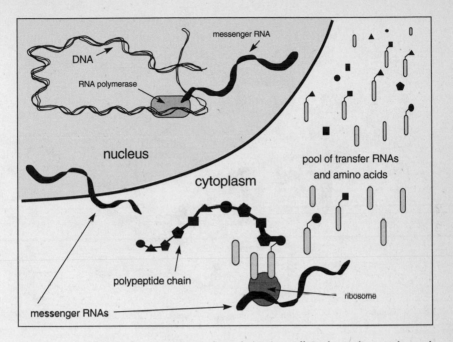

Figure 2.3 Overview of transcription and translation in a cell. In the nucleus, each gene's DNA is transcribed to a messenger RNA under the auspices of the enzyme RNA polymerase and other molecular modulators. The messenger RNA later enters the cytoplasm where it is translated to a polypeptide—a particular chain of amino acids that constitutes a protein subunit ultimately specified by the sequence of nucleotides in the gene that produced the messenger RNA. During translation, transfer RNA molecules (vertical cylinders) pick up and deliver individual amino acids (filled circles and polygons) to a ribosome composed of ribosomal RNAs and proteins. There, the amino acids are hooked together to produce a polypeptide chain.

The enhancers and silencers influence the activity of RNA polymerase indirectly, through connections with large families of activator proteins and repressor proteins that bind with DNA. The regulatory signals from the activators and repressors are further transponded to the RNA polymerase by particular coactivator proteins and basal factors, the accelerator connections and brake lines of our molecular automobile. Altogether, some fifty distinct proteins, each encoded by a different gene, are involved in the process of operating a cell's transcriptional engines. Distinct cliques of these transcriptional factors operate in different types of cells,

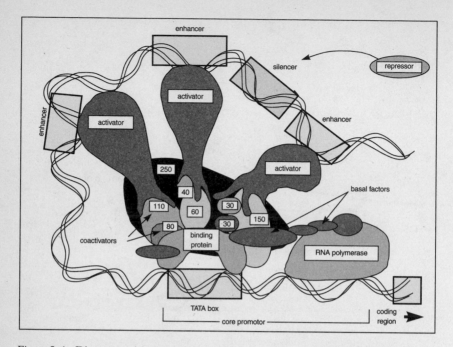

Figure 2.4 Diagrammatic representation of known molecular elements involved in the transcriptional regulation of gene expression. Various proteins physically associated with the meandering DNA strands include RNA polymerase, basal factors, binding proteins, activators, coactivators (numbered according to relative size), and repressors.

and their malfunctions are thought to be responsible for genetic disorders ranging from asthma to various forms of cancers, heart diseases, and immune disorders.

Among the 10,000 or so partially characterized genes recently isolated from various human tissues, nearly 50 percent could be considered to play some regulatory or developmental role (such as cell signaling and communication, control over gene expression, or influences on cell division, structure, or motility). Of course, some regulatory genes have more pervasive and pronounced developmental influences than others. Perhaps the best known of genes with dramatic effects on the developmental control of the body are the homeotic genes. Originally identified in fruit flies, these loci typically encode regulatory proteins that activate or repress other genes through DNA binding. In flies, particular mutations in these genes can cause gross phenotypic changes (such as conversion of antennae into legs, or growth of an extra set of wings where the fly's gyroscopes or "halteres" should be) that are quite like those obtained by the above-mentioned artificial transplantations of imaginal disks. Homeotic genes are suspected to exert basal ontogenetic control by coordinating regulatory patterns in entire batteries of developmental genes.

The structural hallmark of a homeotic gene is the homeobox, a conserved sequence 180 nucleotides long that specifies a DNA-binding polypeptide. Such homeobox sequences recently have been discovered in many eukaryotic organisms. For example, four clusters of homeobox genes involved in limb development are known in vertebrates. Humans have thirty-nine homeobox genes in these four clusters. Recently discovered mutations in one of these genes located on human chromosome 2 produce the conditions of syndactyly and polydactyly: webbed and duplicated fingers. Another homeobox mutation simultaneously produces limb and genital abnormalities.[55] Many other homeotic conditions no doubt are lethal in early human embryos, and hence remain undetected.

In 1764, the leading embryologist of the eighteenth century, Charles Bonnet, proposed that the human egg contained a complete, preformed human in miniature, such that individual development was merely a process of growth. This idea led to the "paradox of emboîtement," in which an embryo within the egg must contain eggs with yet smaller embryos, and so on *reductio ad absurdum*. Attempts even were made to calculate the total number of embryos that could have resided within Eve's ovaries. The modern scientific view, attributable to the discoveries of genetics and cell biology, differs dramatically from such notions.[56] Humans are not preformed or even prefabricated. Rather, each of us is mechanistically created according to instructions contained within an evolved and heritable DNA blueprint. With informational feedback from the zygotic and embryonic environment, the genetic gods choreograph the molecular and cellular dances of development. These miniature ballets are no less beautiful than the most artistic of ceremonial tribal dances designed to celebrate human fertility, or rites of passage from one life stage to the next.

Philosophical Responses to the Scientific Geneses

Scientific studies of biotic geneses have established life's general mode of mechanistic origin and elaboration, as well as human's general temporal place within that evolutionary framework. The timescales involved are so vast as to be nearly incomprehensible to us. To help view the biotic geneses from a more intelligible perspective, consider the following cosmic datebook, which scales evolutionary events into the familiar format of an annual calendar.[57]

Physicists and astronomers who subscribe to the big bang theory of the origin of the universe tell us that it took place roughly 15 billion years ago. If we compress these 15 billion years into one year, the big bang occurs on January 1. The earth comes into existence on approximately September 12, and the first known forms of life appear about October 7. Eukaryotic organisms begin to flourish by mid–November, and by December 17 diverse invertebrate life roams the planet. The first mammals appear on December 26, the first primates on December 29, and the first hominoids on December 30. It is not until late in the final day of our calendar year, December 31 at about 10:30 P.M., that the earliest human creatures amble onto the evolutionary stage. By 11:46 P.M., humans domesticate fire. Thirteen minutes later they are drawing extensive cave paintings, inventing agriculture, and beginning to cluster into the first large towns. Jesus is born four seconds before the present, at 11:59:56 P.M. Within the last second of the cosmic calendar, Europeans discovered the western hemisphere, many countries became mechanized and industrialized, and, perhaps most germane to the current discussion, the experimental method of science, which provides an objective and empirical illumination of the biotic geneses, emerged.

These scientific perspectives are sobering. No longer can we rationally see ourselves as the focal point of creation, either in space or time. We are dwarfed by the immensity of our surroundings, and by the vastness of time within which we participate for so brief an instant. Furthermore, the mechanistic evolutionary processes that brought the genetic gods into existence give every appearance of having operated without proximate external guidance other than that provided by a totally amoral and thoughtless operation, natural selection. Science has illuminated biotic origins at many levels, but also has posed theological predicaments that scarcely could be imagined in prescientific times.

Responses to the recent discoveries in the biological sciences are varied. One approach is a complete denial of evolution. Just as the Catholic church rebuked as heretical Galileo's discoveries in the 1600s, many creationists today consider as blasphemous the findings of evolutionary science. Discussions or arguments with ardent fundamentalists on such matters is often difficult because the epistemological rules of science are flatly incompatible with revelational belief. As noted by Thomas Henry Huxley more than a century ago, "The man of science is the sworn interpreter of nature in the high court of reason. But of what avail is his honest speech, if ignorance is the assessor of the judge, and prejudice the foreman of the jury?"[58]

A second approach is to acknowledge the reality of evolution, but disavow its relevance to the human religious experience. Under this view, evolution is the process by which we came into being but otherwise has no special import for theology or philosophy. It is true that the elucidation of evolution as a mechanistic process dictates no particular religious orientation, any more than Galileo's discovery of the heliocentricity of the solar system dictates particular theological stances. In this important sense, all scientific discoveries can be viewed as irrelevant to theological interpretations of life. Science and religion might be in this way separate realms of human endeavor without any particular significance to one another.

A third approach entertains the possibility of at least some theological relevance for the discoveries of the evolutionary sciences. Under this view, because humans like all other species are products of the evolutionary process, scientists and theologians are justified if not obliged to consider historical contingency as a possible influence on the kinds of religious or philosophical orientations that otherwise might be assumed to be universal, absolute, or of supernatural origin.

THREE

Genetic Maladies

> . . . the germinal cells of different animals, which resemble each other so closely in structure . . . obviously include factors which determine both the forms and metabolic peculiarities of the organisms which originate from them.
>
> Sir Archibald Garrod, Linacre Lecture,
> Cambridge University, 1923

These factors mentioned by Sir Archibald Garrod now are understood to be genes, the corporeal "particles" underlying the particulate inheritance first described by Gregor Mendel. Garrod was a pioneer in the field of human genetics at the turn of the last century, and a prominent early figure in English medicine.[1] Born into a well-to-do physician's family in London in 1857, about the time that Darwin's and Mendel's works first appeared, Garrod was to devote his professional life to the study of human congenital metabolic abnormalities, publishing such works as *Inborn Errors of Metabolism* (1909), and *The Inborn Factors of Inherited Disease* (1931). He now is recognized as the intellectual father of biochemical genetics.

Archibald Garrod elaborated during his career two insightful themes that were apocryphal for the time: first, that due to molecular idiosyncrasies, each person displays a chemical individuality or "diathesis"; and that heritable defects in human metabolism evidence molecular malfunctions, often expressed as enzyme deficiencies in crucial metabolic pathways.

Genetic Individuality

Garrod's conclusion about the biochemical uniqueness of individuals was perhaps the most prophetic. In recent years, laboratory assays of genes and their protein products have provided over-whelming empirical support for Garrod's proposition that each human (barring a monozygotic twin) is distinct from all others in molecular makeup. Thousands of genes have been identified, some with scores of alternative DNA forms (alleles), such that the collective probability of a perfect match between the genomes of any two individuals is vanishingly small.[2] Assays of molecular variation at even small numbers of highly polymorphic genes (usually 5–10 in most courtroom applications, such as the O. J. Simpson trial) produce what are referred to as "DNA fingerprints." Just as conventional fingerprints are person-specific, so too are the genetic "bar codes" now scanned routinely in molecular forensics laboratories to identify the individual source of a blood, semen, or tissue sample.

Nobody supposes that all of these DNA-level variants harm human health. Most of them probably are neutral or nearly neutral with respect to survival and reproduction. Nonetheless, their ubiquity speaks to the concept of human individuality at the molecular level. Furthermore, if even a modest fraction of the variability in the human gene pool is relevant to disease predisposition, then Garrod's concept of diathesis is on firm empirical footing. It may be wise social policy to proclaim that "all men are created equal," but this is incorrect biologically. From a genetic perspective, each person is unique.

Genetic Disorders

One metabolic abnormality studied by Garrod was alkaptonuria, a rare disorder (one in 200,000 births) caused by a defect in the degradative pathway for the amino acids phenylalanine and tyrosine. The condition stems from absence of a genetically encoded enzyme that otherwise catalyzes the breakdown of the intermediary metabolite homogentisic acid (alkapton). As this acid accumulates, it binds irreversibly to the body's collagen (a protein in fibrous connective tissues), producing medical symptoms of degenerative arthritis in the large joints and spine usually beginning in mid-life. Clinical diagnosis is supported by a characteristic darkening of cartilaginous tissues, and the presence of excessive homogentisic acid in the urine that turns black upon exposure to air.

Before Garrod's efforts, alkaptonuria (like many other disorders) was thought to result from a pathogenic organism such as a worm, in this case residing in the kidneys or intestine. Garrod's insight that the disease was caused instead by a chemical error of the human body stemmed from his observations that afflicted infants display homogentisic aciduria within a few hours of birth, and that parents of alkaptonuriac children often were related to each other (e.g., first cousins). Garrod was unaware of Mendelian modes of inheritance at the time of the initial observations in 1901, but his familial case histories soon were interpreted as evidence for genetic transmission according to Mendel's rules.

Although Garrod identified only four inborn errors of metabolism in his original treatise (the others were albinism, cystinuria, and pentosuria), his discoveries were revolutionary. No longer could all human disabilities be attributed to exogenous agents, environmental circumstances, malevolent supernatural Deities, or bad karma. Instead, some afflictions resulted from endogenous, heritable, mechanistically understandable molecular foul-ups.[3] Scientists came to realize that the elaborate metabolic machineries of the human body are subject to heritable design flaws, sometimes with serious health consequences.

With the development and application of biomedical technologies in recent decades, the list of metabolic disorders ascribed to simple genetic defects in humans has grown rapidly (see Figure 3.1). Since the early 1960s, gene catalogues inspired by Garrod's work have provided growing encyclopedias to these human genetic conditions. One of these modern tomes, *The Metabolic and Molecular Bases of Inherited Disease* (MMBID), occupies nearly 5,000 pages of fine print detailing the genetics, biochemical bases, and clinical symptoms of approximately five hundred single-gene disorders in humans. A recent edition of another such compendium, *Mendelian Inheritance in Man* (MIM), describes more than 6,000 human genes, of which about 75 percent are reported to have mutational defects associated with a disease phenotype. In recent years, MIM has been online and is updated daily by computerized searches of the scientific literature.[4]

Some hereditary disorders are characterized far better than others. Alkaptonuria has been studied intensively since Garrod's early

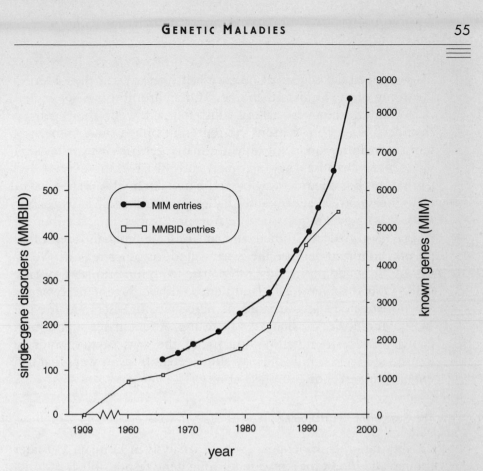

Figure 3.1 Numbers of known human genes and genetic disorders as catalogued in the first eleven bound editions of *Mendelian Inheritance in Man* (MIM) and in the first seven bound editions of *The Metabolic and Molecular Bases of Inherited Disease* (MMBID). Included in the MIM plot is the number of genes catalogued as of May 1997 in the computer online version. The 1909 point refers to the state of knowledge at the time of Garrod's *Inborn Errors of Metabolism*.

research, as evidenced by the sixteen pages devoted to its detailed clinical, biochemical, and genetic description in the 1995 MMBID catalogue. Although a few fine points of the alkaptonuria defect remain unknown, the gene responsible was localized recently to chromosome 3 in the total human complement of twenty-three chromosomal pairs.

Victor McKusick, one of the latter-day giants in the field of human genetic research, has compiled a summary chart of genetically mapped disorders that he ghoulishly refers to as the "Morbid Anatomy of the Human Genome." One chromosome (see Figure 3.2) from this rogues' gallery illustrates a few of these genetic disorders. Such compilations yet represent only a small fraction of genetic malfunctions. Not only are many genetic diseases difficult to identify because they are rare and variable in severity and symptoms, but documenting even the most straightforward hereditary diseases is a lengthy and challenging process. Many genetic disorders are fatal in gametic or embryonic stages, long before they can be identified and studied, and late-onset genetic disorders often go unrecognized because life is truncated for other reasons. Many disorders, including cancers, often arise from mutations in somatic cells rather than germ cells, and hence (although genetic) typically are not included in the ranks of hereditary diseases. Finally, the field of molecular medicine is still young. A reasonable supposition is that particular mutations in any of the tens of thousands of expressed genes in humans may alter metabolisms in ways that can entail some level of clinical disability.

The Search for the HD Gene

To illustrate the challenge of genetic analysis in humans, consider the research quest for a malfunctioning gene responsible for Huntington disease (HD). This inherited malady was named after the young physician George Huntington who in 1872 eloquently described this wrenching condition. HD is a fatal neurological disorder whose symptoms, usually beginning in mid-life, involve uncontrollable movements of the body and progressive dementia. The disease smites nearly ten people in 100,000. In the United States alone, more than 25,000 patients suffer from HD, with about 125,000 more at risk by virtue of being siblings or children of the currently afflicted.

The disease has an interesting global distribution. It is typically associated with populations in western Europe, but it also appears in other geographic hotspots such as Tasmania and Papua New Guinea. The Tasmanian case is understood well. There, a gene for HD can be traced to a widow who in 1848 left her village in Somerset, England and moved to Australia along with her thirteen children.[5] By 1964, descendants of this family accounted for most

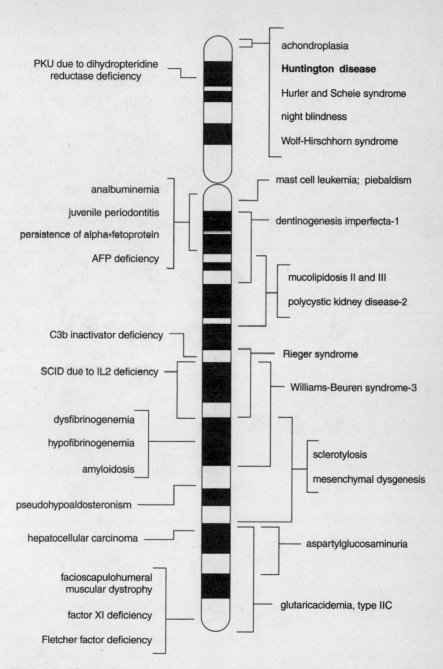

PKU due to dihydropteridine
reductase deficiency

achondroplasia

Huntington disease

Hurler and Scheie syndrome

night blindness

Wolf-Hirschhorn syndrome

analbuminemia

mast cell leukemia; piebaldism

juvenile periodontitis

dentinogenesis imperfecta-1

persistence of alpha-fetoprotein

AFP deficiency

mucolipidosis II and III

polycystic kidney disease-2

C3b inactivator deficiency

Rieger syndrome

SCID due to IL2 deficiency

Williams-Beuren syndrome-3

dysfibrinogenemia

hypofibrinogenemia

sclerotylosis

amyloidosis

mesenchymal dysgenesis

pseudohypoaldosteronism

hepatocellular carcinoma

aspartylglucosaminuria

facioscapulohumeral
muscular dystrophy

factor XI deficiency

glutaricacidemia, type IIC

Fletcher factor deficiency

Figure 3.2 The morbid anatomy of chromosome 4 in humans showing the mapped positions of genes underlying various inherited disorders.

of the 120 afflicted people on the island. In the case of Papua New Guinea, HD probably was introduced by early-twentieth-century whalers from New England, some of whom carried the HD gene. Diaries tell of shipboard visits by "naked and friendly natives," some of whose children inherited copies of the HD gene from sailor fathers. Epidemiologic records indicate that HD has spread mostly through such human migrations from source areas in western Europe.

Examination of the pedigrees of affected families long ago revealed the hereditary basis for HD (see Figure 3.3). A single gene is involved, one defective copy of which inherited by either sex from either parent is sufficient to burden with the disease any

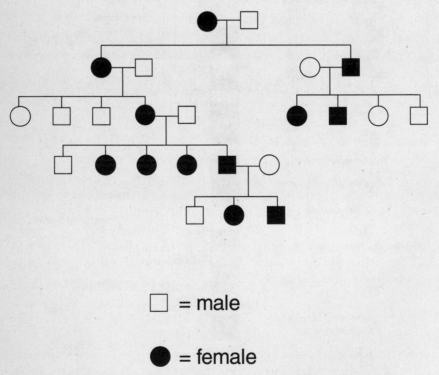

☐ = male

● = female

Figure 3.3 Example of a pedigree for Huntington disease (filled symbols) through five generations of a Venezuelan family at Lake Maracaibo. It is through such family pedigrees that the particular genetic basis of HD first was deduced.

individual who lives long enough. Under Mendel's first law of inheritance, the "law of segregation," each child and full-sib of an affected individual has a 50 percent chance of carrying a copy of the HD allele also, an anguishing prospect for family members who may not yet show symptoms. Mendelian principles have proven so universal that the mode of HD inheritance in humans first was deduced solely from transmission patterns of the disease through family pedigrees.

The highest concentration of HD in the world occurs in isolated villages along the shores of Lake Maracaibo in Venezuela.[6] The disease was introduced (probably by a British sailor) in the early nineteenth century and subsequently rose in frequency to more than seventy times the western European norm. At Lake Maracaibo, medical researchers have administered to the sick, interviewed families and reconstructed their pedigrees, and obtained blood samples from more than 7,000 villagers for molecular genetic analysis in a concerted effort to understand the molecular basis of HD and to find a treatment or cure. Nancy Wexler, president of the Hereditary Disease Foundation and an appointed member of the Human Genome Project, decided to devote her life to the study of hereditary diseases when her own mother was diagnosed with HD. Dr. Wexler has worked tirelessly with scientists and patients in coordinating a scientific assault on the disease.

This attack has taken advantage of the large HD family pedigrees in the Lake Maracaibo area (affected families with a dozen or more children are common) and the availability of numerous polymorphic DNA markers whose locations on various human chromosomes have been determined using techniques of molecular genetics and somatic cell hybridization. The latter laboratory method is especially bizarre. It long has been known that human and mouse somatic cells, when mixed together in a test tube under appropriate culture conditions, spontaneously fuse to form interspecific hybrid cells that initially house a full complement of human and mouse chromosomes. For reasons unknown, human chromosomes tend to be lost more or less at random through successive divisions of these hybrid cells, sometimes until only one remains. By matching the presence versus absence of particular human genes (as determined by molecular genetic assays) against the retention versus loss of human chromosomes in a panel of mouse/human cell lines, which genes belong to which chromosomes can be deduced. Once a battery of such genes is available as a road map to specific chromosomes, the genealogical information in family pedigrees then can help to localize any disease gene of interest.

In the case of HD, the first step in genomic localization involved a search for co-transmission of the disorder with chromosomal marker genes through the large family pedigrees provided by the Venezuelan villagers. Mendel's second law of inheritance, the "law of independent assortment," states that genes on different chromosomes or at distant locations on the same chromosome are not necessarily transmitted together to offspring—they assort independently. The corollary is that genes with nearby addresses on a chromosome usually are transmitted together. Attempts to identify chromosomal markers that displayed co-transmission with HD proved negative for many years, but in 1983 a formerly run-of-the-mill genetic marker (known as G8) on the fourth chromosome yielded the critical breakthrough.[7] By virtue of co-segregation with G8, HD's address was localized to chromosome 4. Further molecular analyses soon honed HD's position to the distal end of the short arm of that chromosome (see Figure 3.2 on p. 57). The exact procedures are too detailed to recount here, except to mention one interesting clue. Another genetic disorder known as the Wolf-Hirschhorn (WH) syndrome (characterized by mental defects and severe retardation of growth) was known to be associated with a visible chromosomal deletion at the distal tip of chromosome 4. In WH patients, the genic region marked by G8 proved to be missing also, thus narrowing the search for the exact chromosomal site for G8 (and HD).

The Huntington disease gene now had a formal chromosomal zip code, 4p16.3, to which researchers could address further genetic inquiries. However, this zip code still encompassed an area of about six million nucleotides of DNA, with the potential for scores of genes, so the address for HD needed much refinement. Initial surveys uncovered several candidate genes, but upon further characterization most were excluded as the actual cause of HD. Finally, in 1993, the HD gene itself was isolated to a location between genes D4S127 and D4S180. On this property in HD patients resides a most malevolent genetic god.

Recent research efforts have identified the structural abnormalities of the HD gene that are the source of the metabolic malfunctions and debilitating symptoms. The HD gene contains a structural motif (of nucleotide triplets CAG, CAG, CAG, . . . CAG) that is correlated with the expression of Huntington disease. On normal copies of the human fourth chromosome, ten to thirty CAG repeats are found in this region of the gene, but the repeat number in HD sufferers is expanded beyond thirty-five copies (up to 120 or more). Huntington disease is one of a growing list of hereditary disorders involving genes that display similar kinds of anomalies in the repeat motifs of tandemly arrayed short sequences.[8]

The molecular search for the HD gene has produced an important clinical byproduct—a predictive diagnostic test for the disease based on laboratory assays of the numbers of CAG repeats. This capability illustrates a general ethical quandary of modern genetic counseling, referred to by Nancy Wexler as the dilemma of Tiresias.[9] In Sophocles's *Oedipus the King,* the blind seer Tiresias confronts Oedipus with the thought: "It is but sorrow to be wise when wisdom profits not." In the current context, this dilemma can be phrased: Given that one of your parents or family members has HD, would you wish to know whether in later life you too will experience this horrible and untreatable brain disease? Thousands of individuals (including Wexler herself) have had to wrestle with this question. Many people understandably decline to take the diagnostic test in the absence of a treatment. But what of those at genetic risk who may contemplate having children? Without knowledge of the test's outcome, a parent at risk for HD (one who has a parent or full-sib already diagnosed with the disease) has a 25 percent chance of transmitting an HD gene to any child. For a prospective parent who tests positive, each contemplated child has a 50 percent probability of inheriting the HD allele. To complicate parenting decisions further, any adult who possesses the HD gene knows that he or she will be permanently disabled in mid-life.

In spite of the many difficulties researchers face, Nancy Wexler chooses to view the study of genetic disorders as affording a challenging, spiritually uplifting opportunity. She speaks in glowing terms of genetic research as "the most ambitious, imaginative, daring effort for humanity to know itself that has ever been attempted . . . it's the best human adventure in the world."[10]

The Chromosomal House of Horrors

To emphasize the troubling scientific and providential enigmas presented by human genetic disorders, and to illustrate the pervasive scope of conditions affected by pernicious genes, I will next describe briefly a few of the more common or gruesome afflictions from the morbid encyclopedia of the human genome. For each, mutations in one or more genes on a human chromosome result in the debilitating diseases mentioned.

Chromosome 1: hypophosphatasia   This genetic defect in skeletal mineralization normally is transmitted as a recessive allele, and can result in symptoms that may include deformed bones and premature loss of deciduous teeth in children. Hypophosphatasia occurs throughout the world, but is notably prevalent in inbred Mennonite families in Manitoba, Canada. There is no established medical treatment.

Chromosome 2: precocious puberty   The dominant allele for this condition is expressed only in males and results in an early onset of testosterone production. Affected boys generally show signs of puberty by the age of four years. This condition exemplifies the profound physiological consequences sometimes resulting from the smallest of genetic alterations. In the entire human genome of 3,000,000,000 nucleotide pairs, this form of precocious puberty is associated with a single nucleotide substitution![11]

Chromosome 3: postanesthetic apnea   The recessive mutation for this condition also involves a single nucleotide substitution that in this case leads to an alteration in nerve impulse transmissions in response to certain chemical stimuli. For homozygous individuals (whose cells possess two defective copies of the gene), a prolonged cessation of breathing may follow administration of a muscle relaxant during surgical anesthesia. Postanesthetic apnea is an example of a genetic condition that may not be a disorder at all under the normal circumstances in which humans evolved, but can become so under a modern environmental challenge. One in about 3,000 North American Caucasians is affected.

Chromosome 5: cri-du-chat syndrome   Named after the unnerving "cry of the cat" wail by afflicted infants, this syndrome is among the most common (one in 50,000 births) of human genetic disorders attributable to a partial chromosomal deletion. The stricken are mentally retarded and have pronounced eye folds, a small face, and a prominent nasal bridge. Other medical complications from the disease often lead to death in infancy or early childhood.

Chromosome 6: Salla disease   This disorder in the body's ability to process and store sialic acid produces noticeable symptoms of poor muscle tone and uncoordinated movements beginning at 6–9 months of age. Approximately one-third of patients never learn to walk, and an equal proportion lose the capacity to produce (but not to comprehend) words. Maturing individuals suffer retarded growth and mental function, and adult IQs are in the range of 20 to 40. Lifespan appears little shortened by the disease, and one man lived to the age of 72. Salla disease is concentrated in northeastern Finland, suggesting that the allele responsible probably traces genealogically to a single mutation that originated in this area.

Several other single-gene diseases have been uncovered in the Finnish population. Finland was colonized only about 2,000 years ago (seventy-five generations), and the population as recently as the late 1600s went through a severe decline. Historically, the Finns have mixed little with other populations, and the country has exceptional family records dating back over three centuries. These factors make the Finnish population a favorable target for molecular genetic studies.

Chromosome 7: cystic fibrosis (CF)   According to northern European folklore, a child who when kissed on the forehead tastes salty is bewitched and soon must die. Excessive sweat is just one manifestation of cystic fibrosis, the most common fatal disorder attributable to an autosomal recessive allele in Caucasian populations (the incidence is one in about 2,500 live births). Thick mucous secretions, often life-threatening, obstruct the lungs of affected children. In a drama as compelling as the quest for the HD gene, another arduous molecular search came to fruition in 1989 with the identification of the offending CF gene, which encodes a protein that channels salt into and out of cells. This gene spans 230,000 nucleotide pairs in the long arm of chromosome 7. A nucleotide deletion that causes the protein product to lack a phenylalanine at position 508 appears to account for about 70 percent of the mutant CF chromosomes worldwide. However, more than five hundred sequence variants in this gene have been discovered, of which at least three hundred and fifty are thought also to produce the pathologic condition.

Chromosome 8: retinitis pigmentosa-1   Retinitis pigmentosa refers to a suite of genetic diseases characterized by degeneration of the eye's retina. First indicated by an inability to see well in poor light, the disease progresses through stages of narrowing tunnel vision to blindness by mid-life. This disorder exemplifies a common situation in which defects in many separate genes can produce similar or identical clinical symptoms, usually because each gene compromises a different step in the biochemical or developmental pathway leading to the disability. Genes implicated in various cases of retinitis pigmentosa have been mapped to chromosomes 3, 6, 7, 8, 11, 14, 16, and the X.

Another genetic disorder recently mapped to chromosome 8 causes individuals to senesce and die early, usually by age fifty. The gene responsible for Werner syndrome encodes a defective DNA helicase enzyme that in normal form appears to play a cellular role in the repair of DNA damages. The mutation leading to Werner syndrome has devastating effects: patients in their thirties typically show pronounced symptoms of old age, such as cataracts, osteoporosis, and heart disease. The Werner syndrome gene provides an unusually clear example of direct genetic control over the aging phenomenon itself.

Chromosome 9: xeroderma pigmentosum-1   This is another disease condition that can result from mutations in many separate genes, one of which is located near the tip of the long arm of chromosome 9. Affected patients show pronounced sensitivity to sunlight resulting in easily parched skin and extreme susceptibility to skin cancers. The median age of children with clinical onset of skin neoplasms is eight years. The disease stems from genetically-based failures in a cell's ability to repair DNA damages from ultraviolet light.

Chromosome 10: porphyria Disorders of porphyrin metabolism provide another example of a condition with a complex etiology that can involve mutations in any of several genes—in this case those involved in the body's ability to produce hemoglobin (the oxygen-carrying molecule in blood). Different forms of porphyria vary in the severity of symptoms, but all tend to be associated with anemia, insomnia, altered consciousness, and intractable pain. King George III, the English monarch during the American Revolution, displayed these symptoms that mystified his doctors but now are appreciated to have stemmed from acute intermittent porphyria, or AIP. The AIP disorder illustrates a general point about gene-environment interactions: Many heritable disorders show variable symptomatic expression as a function of environmental circumstance. Some individuals with the defective AIP gene are asymptomatic throughout their lives. For others, attacks from AIP are intermittent, with debilitating episodes often associated with emotional anxiety or infectious illness. The especially nasty form of porphyria for which a mutated gene on chromosome 10 is responsible produces mutilating skin blisters and scars beginning in childhood.

Chromosome 14: Alzheimer disease This common progressive dementia of the elderly, affecting about four million U.S. citizens alone, is characterized by accumulations of amyloid (starch-like) plaques in the brain. Only 10 to 20 percent of Alzheimer cases are clearly familial, but because of the typical late onset of the disease many inherited cases may go unrecognized. Mutations in several protein-coding genes, notably one encoding an amyloid precursor protein on chromosome 21, are known to contribute to the development of the Alzheimer condition. A form of the disease associated with chromosome 14 shows relatively early onset, often before age sixty. Other genes implicated in familial forms of Alzheimer disease have been mapped to chromosomes 1 and 19, and to mitochondrial DNA.

Chromosome 15: Marfan syndrome This condition first was described in 1896 in a five-year-old girl, Gabrielle, who had disproportionately long limbs, spiderlike fingers (arachnodactyly), tall stature, curvature of the spine, and joint contractures of fingers and knees. Other conditions typically associated with Marfan syndrome include instability of the eye lens, pulmonary difficulties, and susceptibility to hernias. The disease occurs in one out of 10,000 individuals; 15 to 30 percent of these cases represent new mutations. In the late 1980s, molecular detective work identified the culprits, which proved to be mutant alleles of the fibrillin gene located near the middle of chromosome 15.

Chromosome 17: type 1 breast cancer   About 180,000 women in the United States alone are diagnosed with breast cancer every year. Breast cancers have multifaceted etiologies that sometimes include a strong genetic component, as evidenced by the fact that at least 5–10 percent of cases come from families with an obvious history of the disease. In 1994, a BRCA-1 gene that accounts for about one-half of the inherited cases of breast cancer was mapped to chromosome 17. One mutation in this gene is found in relatively high frequency (1 percent) in Ashkenazi Jews whose forebears came from eastern Europe. Its presence increases by more than 80 percent the risk that a woman will develop breast cancer over her lifetime.

Chromosome 19: maple syrup urine disease   This recessive disorder has a pan-ethnic distribution, with a mean worldwide frequency of one per 185,000 infants. The disease gets its name from the characteristic maple syrup odor of the patient's urine, which results from the abnormal accumulation of intermediate compounds from defective steps in the catabolic pathways for particular amino acids. The most severe form of this disorder results in neonatal brain disease and early death. Milder forms can be treated by dietary restrictions on the intake of amino acids that the body cannot process. Screening for maple syrup urine disease currently is conducted in about one-half of the states and a score of other countries.

Chromosome 20: fatal insomnia   Many metabolic disorders are extremely rare. A case in point involved a report of a middle-aged man with sphincter disturbances and severe insomnia. Over the next nine months, the symptoms progressed to dreamlike mind states, tremors, coma, and death. Further inquiry revealed that two sisters of the patient and many relatives over three generations had died from similar symptoms. The disease otherwise was unknown. Fatal familial insomnia (FFI) soon was tracked metabolically to abnormalities of the thalamus portion of the forebrain. The gene responsible resides on chromosome 20, codes for a "prion" protein of uncertain function, and also is implicated in several other genetic diseases of the thalamus.

Chromosome 21: Down syndrome   This genetic disorder involves a gross chromosomal aberration in which a patient carries three copies of a critical portion of chromosome 21, rather than the usual two. The condition lays claim to several firsts: the first chromosomal disorder to be defined clinically (in an 1866 paper by Down titled "Observation on an ethnic classification of idiots"); the first human disorder actually proven to be chromosomal in origin (in 1958); and the highest in frequency of the various forms of mental retardation (one in 700 live births). The physical and physiological hallmarks of Down syndrome include distinctive craniofacial and neurologic anomalies that stem ultimately from metabolic imbalances due to the extra gene copies and their protein products. Prenatal diagnosis via amniocentesis or serum screening is available, and is advised particularly for women thirty-five years and older where the risk to fetuses increases dramatically.

Chromosome 22: DiGeorge syndrome   If a chromosomal duplication can produce medical disorders, it should come as no surprise that the partial or complete loss of a chromosome can do likewise. Cases in point involve DiGeorge syndrome and a related disease attributable to deletions or microdeletions (minimally 300,000 nucleotides long) of a DNA segment on chromosome 22. This disease complex is euphemistically known as CATCH-22. The "22" stands for the chromosomal location, and "CATCH" is an acronym to help physicians remember five hallmark symptoms: cardiac malformations, abnormal facial appearance, thymus gland defects, cleft palate, and hypocalcemia (low calcium levels in the blood).

X chromosome: All chromosomes discussed thus far are autosomes, normally carried in pairs in the diploid somatic cells of both sexes. The full autosomal complement of humans is comprised of forty-four chromosomes total, or twenty-two autosomal pairs. The remaining two chromosomes, X and Y, are the sex chromosomes: Normal human females are XX, males are XY. A haploid egg of a female transmits one X chromosome to each child whereas a father's haploid sperm that fertilizes the egg carries either an X or a Y chromosome with equal probability, thereby deciding junior's gender.

The X chromosome is host to a plethora of genetic defects. For recessive disorders, the deleterious consequences often show higher incidences in males than in females. This is because a single defective copy of an X-chromosome gene in the XY male normally is sufficient to produce the disease whereas the joint occurrence of two defective copies is required for full disease symptoms in XX females.[12] For similar reasons, the incidence of each dominant X-linked disorder is about two times higher in females than in males.

Diseases that are X-linked have characteristic transmission signatures through family pedigrees (see Figure 3.4). For example, because sons receive their X chromosome from mom, X-linked genetic diseases cannot be transmitted from father to son. Furthermore, daughters of affected fathers normally display clinical symptoms only when the defective X-linked allele is dominant (or, if recessive, on those rare occasions when the daughter receives a defective gene copy from mother also). Among the many X-linked diseases involving recessive alleles are particular forms of hemophilia, colorblindness, gout, G6PD deficiency (described later), and Duchenne muscular dystrophy. Vitamin-D-resistant rickets (a condition of soft, easily fractured bones) is an example of an X-linked disorder caused by a dominant allele.

One X-linked inborn error of metabolism, Lesch-Nyhan syndrome, is among the most horrific of all genetic disorders. This recessive ailment is characterized by neurologic dysfunctions that lead to compulsions for vomiting and self-mutilation. Affected children, always boys, exhibit obsessive and uncontrollable urges to harm themselves, for example by chewing away lips and fingers, scalding themselves with hot water, and stabbing faces and eyes with sharp objects. Although mentally retarded, these boys have bright and understanding eyes, feel the pain, and sadly remain aware of their uncontrollable condition. To protect themselves and others, affected children must be restrained physically, from infancy onward. Mothers sometimes are tortured further by guilt when they learn that they transmitted the defective X-linked gene to an affected son.

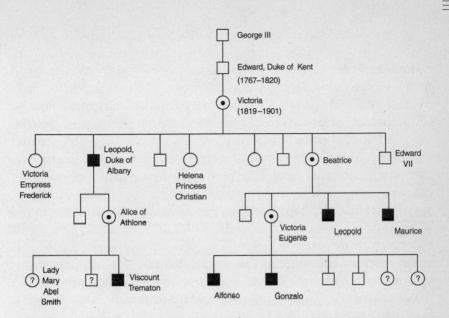

Figure 3.4 A linear, abbreviated pedigree for X-linked hemophilia through European royal families. Males with the disease are shown as filled squares. Queen Victoria of England (the granddaughter of King George III) apparently was the original heterozygous "carrier" (dot inside circle) for the mutant hemophilia allele, and passed the defective copy to several of her children and grandchildren. One daughter, Beatrice, introduced the allele to the Spanish royal family via marriage, as did a son to the Russian royal family. Viscount Trematon and Princes Alfonso and Gonzalo all died following automobile accidents.

In earlier times, children displaying the symptoms of Lesch-Nyhan syndrome were thought to be possessed by demons. Today, we know these demons intimately. They reside within the gene encoding hypoxanthine-guanine phosphoribosyltransferase (HGPRT), an enzyme involved in purine metabolism. The devils themselves usually are point mutations (single nucleotide substitutions) or other minute genetic lesions, the consequences of which are far out of proportion to their size. These genetic demons bedevil more than 2,000 American families alone.

Y chromosome: The Y chromosome is one of the smallest human chromosomes, with "only" 60 million nucleotide pairs. Its male-limited transmission means that any effects of Y-carried genes (of which there are relatively few) are confined to males.

The most fundamental of these effects is sex determination itself.

The gene responsible (originally named testis-determining factor or TDF) was identified recently and shown to be housed on the distal tip of the Y chromosome. Actually, TDF initiates a cascade of events in embryological development that culminates in production of a male. Any environmental or genetic factor that blocks testis differentiation can curtail male formation, leaving female-like "ground states." One such class of genetic defects, XY gonadal dysgenesis, maps to the TDF gene region itself. Affected patients show gradations of sexual ambiguity, ranging from phenotypic males with a micropenis to phenotypic females with a complete absence of male gonads and varying degrees of uterine development and female external genitalia.

The scientific quest for the TDF gene is of interest because it serves to introduce other forms of sex-chromosome anomaly. Early cytogenetic studies uncovered rare instances in which phenotypic males displayed the XX chromosomal constitution normally associated with females. Further analysis showed that these males actually did possess portions of the Y chromosome, but that these had been translocated to the short arm of one of their X's (probably via an abnormal meiotic event in production of their father's sperm). Examination of many such cases led to the identification of the smallest chromosomal transfer producing the XX male condition. This was the small distal tip of the Y. Individuals of XX constitution who possessed other regions of the Y chromosome remained phenotypically female.

Some other common genetic anomalies of sexual differentiation can be mentioned here. Females who carry a single X chromosome (an X0 genotype) display Turner syndrome. Symptoms include short stature, ovarian failure, webbed neck, swollen hands and feet, and constricted aorta. Turner syndrome occurs in an estimated 1–2 percent of all clinically recognized pregnancies, but 99 percent of the affected fetuses die before birth (making Turner syndrome the most common chromosomal anomaly reported in spontaneous abortions). In the general population, the incidence of Turner syndrome is about one per 5,000 female live births. Trisomy for the X (an XXX genotype) is even more common, about one per 1,000 females. The clinical symptoms are relatively mild but often include learning disabilities and partial infertility.

Some common abnormalities in sex chromosome configuration produce individuals who are phenotypically male. These include the XXY (Klinefelter syndrome) and XYY genotypic conditions, both of which occur with incidences of one in about 1,000 male live births. Patients with the former syndrome are tall, thin, and usually infertile; those with the latter show few indications and normally remain undiagnosed.

True hermaphroditism, in which individuals display both testicular and ovarian development simultaneously, also is known in humans. One genetic route to hermaphroditism is XX/XY chimerism, wherein a double fertilization at the time of conception leads to a mixture of XX and XY fetal cells. The individual in effect is a dual embryo composed of two cell types, one genetically ear-marked as male and the other as female. Another route to hermaphroditism involves somatic cell mosaicism. Following the formation of a single fertilized egg that otherwise produces a Klinefelter (XXY) embryo, aberrant separation of sex chromosomes during mitotic cell divisions sometimes results in a mixture of XX and XXY somatic cells, with only the latter leading to testicular development.

Mitochondrial DNA: All genes discussed thus far occur in the cell's nucleus, a sort of command control center demarcated from the cell cytoplasm by a membrane semipermeable to the exchange of cellular products. Outside of this walled compound reside the cell's mitochondria with their own snippets of genetic material (mtDNA). Mitochondria are miniature power plants, generating energy for cellular functions through metabolic processes.[13] Large numbers of mitochondrial power generators exist within each cell.

The cytoplasmic housing for mtDNA has important consequences for hereditary transmission and human diseases. When a tiny sperm and a comparatively huge egg unite, the cytoplasm (and mitochondria within it) in the resulting zygote come predominantly from the egg, and hence from the female parent. In other words, mtDNA is maternally transmitted.[14] Furthermore, unlike the single copy of each nuclear gene that is inherited from each egg (and sperm) cell, large numbers of mitochondria coexist within an egg and are passed to the next generation. Thus, varying mixtures of different mitochondrial alleles sometimes co-inhabit an individual (a condition known as heteroplasmy), and if some of these genotypes prove defective metabolically, any clinical symptoms may vary along a continuum (or, sometimes, across a critical threshold) of severity influenced by the relative proportions of normal and abnormal mtDNA molecules.

Human mtDNA is a closed-circular molecule 16,569 nucleotides in length. Mutations can occur here as well as on the chromosomes in the cell's nucleus, and many are harmful. One example is Leber's hereditary optic neuropathy (LHON), a maternally inherited disease that results in rapid loss of central vision (due to optic nerve death) in young adults. A mutation at one nucleotide position in the mitochondrial ND4 gene causes production of an altered form of an enzyme (ubiquinone oxidoreductase) that leads to the disease. It is both impressive and disheartening that a genetic alteration so trivial can cause such debilitation.

Mitochondrial mutations (including those that arise during the lifetime of an individual) probably play an important role in degenerative disorders of the elderly.[15] MtDNA is minuscule compared to the nuclear genome (less than 0.001 percent as big), but its crucial role in cellular energy production makes it a prime candidate for age-related dysfunctions. An accumulation of mtDNA damages in somatic cells may account in part for gradual declines in the cellular capacity to generate energy. These mtDNA damages often result from mutagenic saboteurs known as free radical molecules that occur in unusually high concentrations in mitochondria.[16] Tissues and organ systems most affected by energy brownouts are those with high energy demands, such as the central nervous system, heart, skeletal muscle, pancreas, kidney, and liver. Mutations in nuclear genes are involved too, because many of the enzymes in the energy-generating pathways are nuclear encoded and imported into the mitochondrion where they interact with the mitochondrial gene products.

Genetic Disorders with Complex Etiology

Many of the human genetic diseases discussed thus far stem from specifiable mutations at single genes. Though common, disorders with such simple genetic basis are the exception.[17] Most clinical metabolic disabilities have multifactorial origins that involve inter-actions among multiple genes (polygenes) often in conjunction with numerous environmental influences. Such complex diseases include diabetes, rheumatoid arthritis, and hypertension, to name just three. The conditions briefly described next—cancer and cardiovascular disease—are the two current leading causes of death in the U.S. population.

Cancer Cancers come in many awful varieties, but all involve an uninhibited proliferation of renegade somatic cells that fail to obey the normal constraints of regulated cell division. Each incidence of cancer traces to a single somatic cell whose clonal progeny develop into cellular masses that can damage adjacent tissues or metastasize to proliferate at other sites in the body. How are otherwise healthy cells of the liver, heart, lung, brain, skin, colon, prostate, uterus, or breast converted to these malignant internal parasites? Current research in molecular genetics provides some answers.[18]

In the progression from cellular normalcy to cancer, genetic changes cause a cell to lose its capacity to respond appropriately to the stimulatory and inhibitory signals that otherwise govern cell division. Sometimes one dominant mutation (e.g., in a gene specifying a protein that encourages cellular replication) suffices to induce cancerous growth, but more often a succession of mutations at multiple genes is involved. This mutational progression may be facilitated when one of the initial genetic lesions impairs a cell's capacity to mend environmentally induced damages to DNA,[19] or when an individual has been exposed to environmental carcinogens during his or her lifetime.

The following "clonal evolution" scenario summarizes modern thought about the usual emergence of cancer. A normal somatic cell may have inherited or acquired (e.g., from a virus)[20] a mutation that disposes it genetically toward less-regulated cell division. The mutation is passed to cellular progeny of the dividing cell. Eventually, one of these cells mutates at a second site, one of its cellular offspring mutates at a third site, and so on. Eventually, some descendant cell happens to accumulate a sufficient number of particular mutations (perhaps 5–10) to cross the threshold into a full-blown malignant cancerous form.

This general etiological model, many molecular details of which have been worked out for several cancers including those of the brain and colon, accounts both for the polygenic nature of many cancers, and for the observation that many human families are cancer-prone. A particular mutation that biases toward cancer can be transmitted not only through somatic cell lineages within an individual, but also in some cases across generations through a family's germ lines. Inheritance of such mutations predisposes a person to cancer, but whether the cancer materializes depends on the idiosyncratic occurrence of further mutations that may arise during the individual's lifetime.

Cardiovascular disease   Overt malformations of the blood circulatory system occur in about 1 percent of live births and account for perhaps ten times more still births. Elevated arterial blood pressure (hypertension) affects some 15–20 percent of adults in many industrialized societies. In the United States, cardiac arrhythmias are responsible for 250,000 deaths each year, with cardiomyopathies and atherosclerotic vascular diseases (primarily coronary artery disease and strokes) adding another 25,000 and 600,000 deaths, respectively. The worldwide toll ranks the cardiovascular diseases among the leading killers and debilitators of our species.

Intensive medical research in recent decades has begun to illuminate the complex genetic bases of cardiovascular disease.[21] Formation of the heart entails intricate interactions among cells with multiple embryonic origins, and the involvement of multiple genetic pathways. Mutations in the genes of these pathways can result in abnormalities of cardiac morphogenesis, as in the DiGeorge syndrome discussed earlier. With respect to hypertension, at least ten genes have been shown to affect blood pressure through a common pathway influencing salt and water reabsorption by the kidney. Cardiac arrhythmias, cardiomyopathies, and vascular diseases have been linked to a variety of genetic defects underlying the production of ion channels in cells, contractile and structural proteins, and signaling molecules. In general, cardiovascular diseases tend to display complex multifactorial genetic bases, and in addition are influenced greatly by environmental factors such as diet and stress that can push genetically predisposed individuals into the realm of clinical symptoms.

Infectious Disease

To anyone who remains unconvinced of the pervasive influence of genes on human health, consider also the connection to infectious microbial parasites. A multitude of infectious viruses, bacteria, and yeasts provoke colds, influenzas, and many other annoying and sometimes life-threatening human diseases. Protozoans that produce malaria, sleeping sickness, Chagas' disease, leishmaniases, and other diseases infect more that 10 percent of the world's population and account for tens of millions of deaths every year. Throughout human history, epidemics such as smallpox have touched the lives of countless millions of people. The virus responsible for smallpox recently was eradicated,[22] and the polio virus too may soon meet its demise, but numerous other microbes remain to cause such age-old scourges as measles, cholera, pneumonia, hepatitis, and tuberculosis.

Infectious microbes contain their own genes that impact human and microbial affairs, and science has unveiled many of the modes of their actions. As noted by Lewis Thomas twenty years ago, "Without the long, painstaking research on the tubercle bacillus, we would still be thinking that tuberculosis was due to night air, and we would still be trying to cure it by sunlight."[23] One mounting problem in the quest to combat infectious diseases is that microbes often evolve resistance to vaccines, drugs, or other therapies directed against them. Tuberculosis bacteria, for example, have become increasingly resistant to standard antibiotics, as have the microbes responsible for malaria, cholera, pneumonia, blood infections, and a host of other infectious diseases.

On the human side of the equation, people vary greatly in their genetic susceptibility to infection. One dramatic example is provided by human resistance to the AIDS virus. Doctors long have been intrigued by the fact that some individuals repeatedly exposed to HIV nonetheless remain free from the disease. There is now an explanation for this phenomenon.[24] On the cell surfaces of HIV-susceptible individuals, a glycoprotein molecule called "CC-chemokine receptor 5" mediates cellular entry of the virus. In HIV-resistant individuals, a truncated form of the glycoprotein (due to a deletion mutation in the CCR-5 gene) blocks HIV entry. The frequency of this mutant allele in the Caucasian population is about 10 percent, and natural resistance to infection is strongest in homozygous individuals.

Enigmas Posed by Malevolent Genes

The collective burden that salient genetic defects place on the human population is referred to as genetic load, and many quantitative attempts have been made to weigh this millstone. Representative summaries indicate that chromosomal abnormalities associated with medical ailments are displayed by about 1 percent of the human population, known single-gene disorders by another 1 percent, congenital genetic malformations by 2 percent, and other overt disorders with an evident genetic component by an additional 1 percent. Thus, at least 5 percent of the human population is known to be afflicted with obvious genetic disabilities. About 20 percent of infant deaths are attributable to genetic defects, as are nearly 50 percent of pediatric and adult admissions to hospitals. Apart from the tremendous burden to affected individuals, the costs to society are staggering.[25]

Although medically significant, these tabulations provide only a minimum estimate of genetic load because of serious downward recording biases. First, most of these figures exclude the vast majority of diseases for which hereditary components are suspected but ill-defined, including circulatory disorders, various cancers, and the multitudinous disorders of normal aging. These conditions typically have multifactorial genetic (and environmental) involvement, but the actions of individual genes in the complex nexus of causality can be difficult to pinpoint. Second, the figures exclude ubiquitous genetic defects that strictly speaking are not heritable because they are confined to somatic as opposed to germ cells. Many cancers qualify. Third, the figures fail to include genetic defects that go undetected because they merely lower fertility, or because they terminate life well before birth, yet these are precisely the biological arenas where the most serious selective action against compromised genotypes takes place. For example, chromosomal defects alone occur in about 20 percent of pregnancies, and genetic flaws appear to be responsible for more than 50 percent of all miscarriages.

The genetic gods produce a vast number of inborn errors of metabolism. This documented truth raises troubling epistemological issues for both religious (providential) and scientific discussion.

From a religious perspective, genetic malfunctions pose a number of ethical enigmas. Why would an omnipotent and loving god cut life terribly short, prolong suffering over decades, devise hideous self-torturing behavior, or send disease to only certain age groups or ethnic groups? How could a benevolent god countenance such horrific human suffering? Humans through the ages have proposed any number of reasons. Perhaps a god does not exist, or is less than all-powerful. Maybe a god possesses supreme powers but fails to exercise them. Perhaps a god purposefully designs genetic errors as a test of the afflicted's faith, or as finely tuned damnations for infractions of his will. However, the astonishing severity of many of the punishments, and their apparent quirkiness of allocation, fail to fit the crimes. Indeed, genetic punishment frequently is meted out to those normally perceived as most innocent: unborn fetuses, and the aged or already infirm. Perhaps an omnipotent god's concepts of love, fairness, suffering, and morality all differ fundamentally from the usual meanings of these words to most of us.

Another class of providential explanation is that sufferings in this life are spiritual tolls for crimes committed in past lives. In quite a different sense, this explanation has an element of scientific truth. Many genetic defects in the present human population were inherited from our ancestors, and not generated *de novo* in the current generation. The genes in our hereditary blueprints do have past lives that can haunt us.

Inborn errors of metabolism pose profound explanatory challenges to scientific beliefs as well as to religious beliefs. Why do genetic mutations detrimental to fitness persist in human populations? Why hasn't natural selection's concern with reproductive performance eliminated the human suffering that surely has negative impacts on survival and reproduction? The scientific answers are clearer. Inborn errors of metabolism exist not because of the malfunctions they produce, but despite them.

Many of the rarer inborn metabolic disorders are encoded by mutations not yet eliminated by natural selection, either because the mutations are partially camouflaged, or because their harmful effects are modest in relation to the rate at which the mutations arise. Camouflaging can occur in at least three ways. Alleles responsible for many genetic diseases, such as alkaptonuria, have deleterious consequences only in homozygous individuals. In heterozygotes, these alleles are shielded from natural selection's view because their poor metabolic performance is compensated by the normal allele. Also, many genetic disorders such as Alzheimer disease have postreproductive onset. Contemporary natural selection is blinkered from the scrutiny of genetic defects whose consequences are postponed beyond reproductive age because these defects normally fail to lower an individual's reproductive fitness. In a sense, senescence and death themselves are inborn genetic diseases, and an important evolutionary question is how to account for the ubiquitous occurrence of these phenomena. A third form of evolutionary camouflaging arises because deleterious effects of some mutant alleles are evident only in some environments. Phenylketonuria is an autosomal recessive disorder characterized by severe mental retardation due to the accumulation of phenylalanine and related metabolites in the body. A mutation that knocks out the function of the enzyme phenylalanine hydroxylase is responsible. However, if the condition is diagnosed in early infancy, a diet low in phenylalanine can compensate for the enzyme's inactivity to the extent that some patients achieve normal intelligence.

Many genetic disorders typically are not discussed as disorders at all because their harmful expression is confined to environments viewed as aberrant. For example, our genetic inability to produce vitamin C is of no health consequence when ascorbic acid from fresh fruits and vegetables is available. However, scurvy results when dietary access to vitamin C is limited, as often was true for European sailors on prolonged voyages during the fifteenth to nineteenth centuries. Conversely, some genetic conditions such as postanesthetic apnea and many disorders of the elderly have become increasingly visible in our modern environment.

In conjunction with genetic, developmental, and environmental camouflaging, recurrent mutation also contributes to the maintenance of deleterious alleles in human populations. The theory of mathematical population genetics shows that the expected frequency of a harmful allele at any gene is influenced by a balance between the origination rate of that allele by mutation (m) and the selection-mediated loss of the allele due to its fitness-reducing effects. Over the long term, the tug of war between the forces of recurrent mutation and purifying selection tends toward an equilibrium population frequency for the deleterious allele.[26] These allele frequencies are low but nonzero for realistic mutation rates (which are typically 10^{-5} or lower for point mutations per gene per generation). The balance achieved between deleterious mutations and cleansing natural selection accounts for the observed frequencies of many rare genetic disorders.

For example, a recessive lethal allele that arises at mutation rate $m = 10^{-5}$ achieves a mutation-selection balance at a population frequency of about $q = 0.0032$. Death results when two copies of the defective allele appear together in an individual, an occurrence expected with probability $q^2 = 10^{-5}$. In other words, in a population of size 1,000,000, about ten people per generation are expected to die from this hereditary disorder, a rather typical figure for many serious genetic diseases. Some gross chromosomal disfigurations, such as loss of the Y or the presence of three copies of chromosome 21, occur spontaneously at higher frequency (e.g., $m \simeq 10^{-2} - 10^{-3}$), thus accounting for the higher incidences in human populations of genetic disorders such as Turner and Down syndromes.

In some cases, natural selection itself acts in a manner that maintains high frequencies of genotypes that at first examination appear deleterious. An example involves the most common human enzymopathy known: glucose-6-phosphate dehydrogenase (G6PD) deficiency, which affects more that 400 million people worldwide. This genetic condition, inherited on the X chromosome, can result in severe hemolytic anemia following an infection, ingestion of certain drugs, or consumption of particular foods.

Why hasn't natural selection scrupulously culled such deleterious alleles from the human gene pool, driving them to a low frequency balanced only by recurrent mutation? The answer appears to be that the G6PD deficiency simultaneously confers upon its bearers a startling reduction (46–58 percent) in susceptibility to malaria, an evolutionary benefit that has compensated for the cost of the deficiency.[27]

Another genetic polymorphism related to malarial resistance involves a hemoglobin gene. Under low oxygen conditions, the red blood cells of individuals homozygous for the "S" allele assume a rigid configuration, clog blood capillaries, and produce a painful and life-threatening sickle cell disease. The S allele reaches frequencies of 20 percent in some African populations, far higher than anticipated from recurrent mutation alone. An explanation long has been known. The S allele has attained high frequency because it also affords heterozygous individuals an increased resistance to malaria. Thus, in malarial regions, heterozygotes have a fitness advantage over normal (A/A) homozygotes, and they also have an advantage over S/S homozygotes by virtue of a near freedom from sickle cell disease. Natural selection operates so as to retain both alleles in frequencies determined by the relative fitnesses of the two homozygous classes.[28] Under the rules of Mendelian inheritance and sexual reproduction, heterozygotes do not automatically pass these advantages directly to their offspring, and in each generation new homozygotes are produced. This produces a segregational load that contributes to the total genetic burden that humans bear.

In summary, mutational and selective influences provide the proximate scientific explanations for why humans are burdened with inborn errors of metabolism. These processes are oblivious to pain and suffering—they are both mindless and amoral. But why do these natural evolutionary processes themselves exist? Why do harmful mutations arise? Why are they shuffled and redistributed through sexual reproduction in a seemingly random fashion? How can the genetic gods (or any other gods) play such games of dice with our lives? As we shall see, science has provisional answers to these questions as well.

FOUR

Genetic Beneficence

[Cleanthes] . . . [T]he Author of Nature is somewhat similar to
the mind of man, though possessed of much larger faculties,
proportioned to the grandeur of the work which he has
executed . . . [B]y this argument alone, do we prove at once the
existence of a Deity.

[Philo] . . . [W]hat surprise must we entertain, when we find
him a stupid mechanic.

David Hume, *Dialogues Concerning Natural Religion* (1779)

The "argument from design" advanced by Hume's fictional
character Cleanthes summarizes traditional logic underlying
claims for perfection in the omnipotent forces that have produced life. The rejoinder by Cleanthes's friend Philo identifies a
fundamental problem in the argument from design, a difficulty that
has only gained force with recent molecular findings from the
evolutionary-genetic sciences. This chapter will explain why.

Genes provide both the factual basis and the evolutionary reason
for our existence. These heavy responsibilities seldom stem from
the direct action of DNA itself, but instead from its coding properties. Much of the hereditary blueprint is transcribed by cellular
processes to ribonucleic acid (RNA) molecules, each of which in
effect is a structural mirror image of one strand in a portion of the
original DNA archetype. Structural genes code for messenger
RNAs (mRNAs), which in turn are translated into myriad proteins
that contribute to the physical fabric of life, or that serve as enzymatic catalysts for life's biochemical reactions. Other genes code
for ribosomal (r) and transfer (t) RNAs that assist in protein translation. Some DNA units, called regulatory sequences, modulate

81

the activities of structural genes or of other regulatory genes—for example by overseeing fundamental cellular activities such as DNA replication, transcription, and translation. Other stretches of DNA do little more than contribute to the general physical framework of chromosomes. Some ultra-selfish stretches of DNA have no known function other than their own self-perpetuation.

One of the scientific highlights of the twentieth century was the discovery in 1953 by James Watson and Francis Crick of the chemical composition of DNA.[1] Making up this simply elegant molecule are two complementary strings of nucleotides intimately intertwined like two mating cobras. DNA's double-helical architecture at once suggested the molecule's dual role: as a blueprint for life, and as a template for replication. In blueprint mode, DNA contains all of the instructions necessary for the construction and functional operation of a human being. In photocopy mode, the redundancy inherent in DNA's double helix provides a straightforward mechanism for molecular self-reproduction. During DNA replication, the complementary strands unzip and each strand provides a template for the reconstitution of its Siamese twin (see Figure 4.1).

The incredible volume of information within any genome stems not from a great diversity among DNA's chemical constituents, of which there are only four types, but rather from the vast numbers of arrangements possible when these four classes of nucleotide subunits are organized variously into long, linear strings. The human genome consists of some three billion nucleotide base pairs (bp), more than a hundred times the total number of letter characters in an eighteen-volume World Book encyclopedia. With respect to information storage, the binary code of digital computers and Morse code function in much the same manner as DNA code, but with only two coding units (plusses and minuses, dots and dashes, respectively) rather than four.

Recent research has capitalized upon the digital power of DNA to solve calculation-intense mathematical problems in the laboratory. In 1994, researchers employed a "DNA computer" to solve a classic computational problem in which a "traveling salesman" must determine the shortest total path to visit numerous towns.[2] In a test tube, each town was represented by a random DNA strand

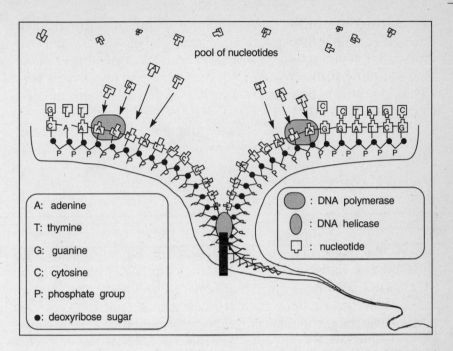

Figure 4.1 During DNA replication, helicase enzymes unzip double-stranded DNA and each single strand serves as a template for the synthesis of a complementary strand from the pool of nucleotides available in the cell nucleus. The process is catalyzed by DNA polymerase and by a host of other molecules.

of twenty nucleotides. Each route was represented by another DNA strand of the same length, the first ten nucleotides matching particular towns of origin, and the last ten matching destination towns. Huge numbers of these DNA strands were mixed together and permitted opportunities to join in any conceivable way. From out of this DNA broth, the scientists then used standard biochemical tools to retrieve strands from the starting and ending towns, and all connecting stops that had become joined by base-pair matches. These matches described the route the traveling salesman should take. The power of this approach stems from the fact that even small test tubes hold trillions of DNA copies that act like parallel mega-processors, performing multitudinous (10^{18}–10^{20}) computational operations at once. It is humbling to realize that organic evolution's biochemical devices for information management can be harnessed to perform certain mathematical tasks that rival or even surpass the capabilities of the fastest inorganic computers.[3]

In the DNA code as it resides within real organisms, long sequences of the four types of nucleotides carry the evolved biochemical messages necessary (but not alone sufficient) for life. Nearly every human cell carries this entire genetic script within its chromosomal chapters. This human calligraphy differs in detail, but not in general features or style, from the genetic encyclopedias specifying the production and operation of armadillos, whooping cranes, or worms.

Inborn Gifts of Metabolism

The genetic gods may take away our health, but so too can they provide. For many of us, especially in our youth, health is an accustomed state interrupted occasionally by illnesses or injuries that send us scurrying to doctors, shamans, or faith healers. This genetic "birthright"—among the most wonderful of gifts—must seem terribly unfair to those to whom it has been denied and who must struggle with disabilities that the rest of us may never experience.

The mere fact that I am able to type this sentence (and you to read and comprehend it) is itself a testament to the beneficent side of genes. My eyes (albeit with the assistance of reading glasses) pick up the varying wavelengths of light from the computer monitor and transmit them via nerve impulses to my brain, which at the moment mercifully is relieved from its usual writer's block. This translation appears to me not only as visual images, but also as intelligible words and ideas. In a feat that surpasses the capabilities of current inorganic computers, I am cognizant of this paragraph's contents. As you read these lines, equally miraculous chains of biotic events in your nervous system similarly convert pages of text into visual images and cognition, colored and filtered by your own particular experiences.

Some elements of sensual awareness, such as color perception, are known to have a relatively straightforward genetic foundation. Eye pigment proteins sensitive to different wavelengths of light are encoded by several tandemly aligned X-chromosome genes, mutations in which can produce various forms of "color blindness." In the late 1700s, the famous chemist and physicist John Dalton became aware of his color blindness when colleagues' descriptions of polarized bands of light from a prism failed to match his own. In a worthy example of scientific commitment, Dalton instructed that following his death his eyes be saved for posterity so that someday they might be studied to assess the biological basis for the condition. Dalton's dream came true nearly two centuries later, in 1995, when molecular geneticists used the "PCR" technique of recombinant DNA technology to assay bits of eye tissue that had been preserved by Dalton's medical attendant. The modern genetic assays revealed that one of Dalton's eye pigment genes had been deleted by a mutation, leading to a condition known as deuteranopia: an inability to distinguish colors in the red end of the spectrum.[4] A single gene deletion altered Dalton's basic perception of the world.

Our physical and athletic attributes are no less remarkable. Consider the elemental task of walking. In coordinated fashion, myriad nerves, tendons, and muscles rhythmically operate to propel us forward, maintain balance, and avoid obstacles. Such complex physical skills are all the more amazing in light of the fact that all functions and forms of each human body are outcomes of a developmental process that traces back just a few years earlier to a single fertilized egg cell. Notwithstanding the importance of environmental influences, this ontogeny is fundamentally a genetically driven progression.

Meanwhile, our bodies quietly go about their routine but hardly mundane internal housekeeping functions. Our digestive systems convert food to energy. Our immune systems patrol for unwanted microbial invaders. Our DNA repair systems mend most of the thousands of DNA damages from ultraviolet light and other environmental insults that each cell receives every hour. Our nervous systems monitor sights, smells, and sounds of danger. Our endocrine systems prepare us for sex, combat, or a racquetball match, which will tax our respiratory, circulatory, and musculoskeletal systems more rigorously than usual. In general, the 100 trillion cells that make up a human body go about their molecular tasks every passing moment until we die.

The molecular events underlying these tasks are the business of genetically encoded cellular processes. To illustrate the sheer complexity of the mechanistic operations, consider a small subset of the molecular action within one cellular location—the mitochondrion. Figure 4.2 provides an abbreviated sketch of a portion of the mitochondrial pathways by which ingested sugars (carbohydrates) and fats (fatty acids) are broken down to fuel cellular energy production. Following a meal, sugars initially are processed by a multistep glycolytic pathway (outside the mitochondrion) into molecules of pyruvate that then are transported across the outer mitochondrial membrane and converted to molecules of acetyl-

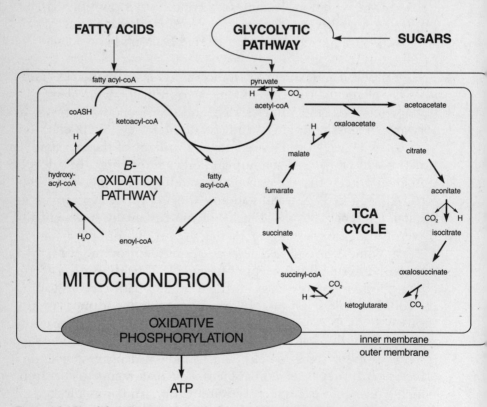

Figure 4.2 Simplified diagram of some of the biochemical pathways within mitochondria by which organic sugars and fatty acids are processed toward production of the cell's energy currency, ATP.

coA. These then enter the depicted tricarboxylic acid (TCA) cycle. Other acetyl-coA molecules from the β-oxidation degradative pathway of fatty acids join this metabolic merry-go-round. At each and every step of the operation, gene-encoded enzymes[5] facilitate the biochemical conversions that otherwise would come to an abrupt halt. For example, the alteration of malate to oxaloacetate in the TCA cycle is catalyzed by malic dehydrogenase, an enzyme encoded by a gene on human chromosome 2.

Several steps in the catabolic cycles of TCA and β-oxidation result in the release of hydrogen and carbon dioxide from organic foodstuff. The hydrogen then is burned with oxygen, and the energy released charges the mitochondrial membrane, thereby creating a capacitor that helps drive the synthesis of adenosine triphosphate (ATP). This ATP synthesis takes place on the mitochondrial inner membrane in a complicated molecular process known as oxidative phosphorylation, conducted under the auspices of numerous structural and enzymatic proteins. As a primary marketable product of the cell's mitochondrial power plants, ATP is the universal energy base for numerous cellular functions: the biochemical equivalent of the electricity harnessed to drive your kitchen's toaster and refrigerator.

There are many such biochemical pathways at work in every cell, busily processing amino acids (the building blocks of proteins), other organic acids, purines and pyrimidines (the building blocks of DNA), proteins and lipoproteins, antibodies, hormones, and the host of other organic molecules necessary for life. In each pathway, gene-encoded enzymes and structural proteins (such as those that contribute to membranes and other physical features of the cell) play critical roles, as do regulatory genes that help to direct cellular operations. Altogether, thousands of genes contribute to the mechanistic underpinnings of human health. The immediate point here is not to belabor the nuances of molecular events and pathways themselves, but rather to call attention to the richness and expertise of the molecular orchestra, and the beauty of its resulting symphony.

Ironically, genes make themselves known to us most clearly when they malfunction. In our personal and family lives, we become acutely aware of genetic influences when genes and their products fail to perform properly. In scientific research too, genes usually are first identified and studied when the processes they govern go awry. For example, Sir Archibald Garrod was able to identify a conditional defect in the metabolism of homogentisic acid only because some individuals display the dark urine and other symptoms of alkaptonuria, and it was only via the transmission pattern of this aberrant condition through family pedigrees that the Mendelian basis of the disorder was deduced. Had no assayable variation in the genetic blueprint for homogentisic acid oxidase been available, the mere existence of this enzyme-coding gene would have gone unrecognized, particularly in Garrod's time.[6]

How many human genes exist? The minimum number can be no less than the approximately 10,000 functional genes already identified.[7] The maximum number can be no greater than would fit within the structural confines of the three-billion-base-pair human genome: about 150,000 genes, under a reasonable supposition that an average gene spans roughly twenty kilobases (20,000 bp).[8] However, there are reasons to believe that the true number of functional genes contributing to human health is probably considerably less than 150,000.

One line of evidence is that about 30 percent of the human genome exists as families of short, reiterated DNA sequences, many of which are nontranscribed and as yet have no well-documented function. For example, one class of highly repetitive DNA sequences (the alpha family, with a basic repeat unit of 171 bp) occurs in tandem arrays up to several million base pairs long.[9] A shorter, tandemly reiterated sequence of only 6 bp (TTAGGG) is found at the ends of every human chromosome, in arrays of up to 5,000–10,000 copies. Another well-known class of highly reiterated sequences (the Alu family) consists of several hundred thousand copies of a 300 bp sequence scattered around the human genome at many sites. Alu sequences sometimes are transcribed, and some of them may compose "transposable elements," but in any event their functional contributions to human health and well-being remain problematic. On average, the various families of highly reiterated DNA have a repetition frequency of about 50,000 copies per genome! Whether these are to be included in the tally of genes is definitional, but if they are, the structural redundancy of some genes must be appreciated.

"Microsatellite" regions (see Chapter 3, note 8) consist of modest numbers of tandem repeats of simple DNA sequences typically two to four nucleotides long. These too are ubiquitous in humans and other species. For example, a recent map of the human genome identified 5,264 different chromosomal positions for a tandem di-nucleotide repeat unit ACACACAC . . . AC.[10] These repetitious DNA sequences, presumably often functionless themselves, nonetheless are extremely useful to medical researchers as polymorphic markers that facilitate identification of operational genes at chromosomal sites that are sometimes adjacent.

Other reiterated sequences exist as families of "middle-repetitive" DNA, usually present in tens to hundreds of copies each per genome. The best known of these sequences encode the large and small subunits of ribosomal RNA molecules that are needed in great abundance by cells for translating messenger RNAs into proteins. These rRNA genes are arranged tandemly in clusters known as "nucleolus-organizing regions" along the short arms of chromosomes 1, 13, 14, 15, 21, and 22. Other middle-repetitive sequences include approximately sixty different families of transfer RNA genes, each with about ten to twenty copies sometimes widely dispersed throughout the genome. Not all middle-repetitive families have such housekeeping functions. Some exist as active or silenced transposable elements, which are abundant in plant and animal genomes and often appear to behave in their own rather than their hosts' interests.

"Pseudogenes" constitute a second category of DNA sequences without clear functional relevance. These now-silent stretches of DNA bear clear compositional resemblance to functional genes, and indeed originated from them, but subsequent mutations have garbled their meaning such that they no longer specify an operational gene product.[11] For example, several hemoglobin pseudogenes differ from their functional counterparts by well-characterized mutations that render them useless for globin production. "Classical" pseudogenes, often with chromosomal addresses near their progenitors, arise through regionalized gene duplications. "Processed" pseudogenes arise from genomic insertions of extra DNA copies through intermediate RNA molecules, and, hence, can appear anywhere in the genome. In general, both types of nonfunctional genetic relics are common if not pervasive in nuclear genomes.

"Introns" provide a third reason for concluding that much of the human genome consists of DNA sequences that are perhaps unnecessary or at least functionally rather indolent. These "intragenic regions" are stretches of nucleotides transcribed to mRNA, but not subsequently translated to protein. Occurring as spacers within structural genes, introns separate a gene's exons (expressed regions) that actually encode protein subunits. Introns can vary in size from about 100 bp to more than 100,000 bp, and dozens of introns exist within some loci, often accounting for more than 90 percent of the total length of a structural gene. Walter Gilbert, who coined the word intron in 1978, put it thus: "The gene is a mosaic: expressed sequences held in a matrix of silent DNA, an intronic matrix."[12] Although introns can play important roles with respect to gene organization and expression, their often rapid pace of evolution suggests that the particular sequences therein are relatively unconstrained.

In terms of total length and general organizational features, the human genome is unremarkable when compared to those of other mammals and most nonmammalian vertebrates. Thus, whatever makes a human a human, a bird a bird, and a fish a fish, cannot be decided by considerations of genomic size or sequence complexity alone.[13] Indeed, the genomes of a few fish and amphibian species are up to fifty times larger than that of humans. At the other end of the scale, the smallest genome yet reported for any vertebrate belongs to a pufferfish. Its genome is a "mere" 400,000,000 bp in length, about 7.5 times smaller than that of humans.

All vertebrate animals possess the same basic suites of cellular and metabolic pathways, so any tally of pufferfish genes should provide a good first approximation of the number of functional genes in humans also. The compact nature of the pufferfish genome stems in part from a relative paucity of repetitive sequences, which constitute only 7.4 percent of the total genomic length. Thus, all of the pufferfish's functional genes must fit within the confines of

370,000,000 bp,[14] enough room to accommodate about 18,500 loci if a typical structural gene spans 20,000 bp. However, pufferfish introns appear to be smaller than those of humans, meaning that the average pufferfish gene might be shorter too. Thus, somewhat more than 18,500 genes may reside within its (and our) genomes.

Taking all considerations of genomic size, genetic redundancy, and nonfunctionality into account, the actual number of genes making positive contributions to human health is probably on the order of 50,000 to 100,000.[15] Within the next decade, such rough estimates will become moot as the true gene number is determined directly and more precisely from findings of the Human Genome Project. In any event, whether such numbers are to be interpreted as large, moderate, or small is a matter of taste. An astonishing degree of genetic and metabolic complexity is indicated by these polytheistic tallies, but on the other hand, the number of genetic gods is hardly so vast as to be unfathomable to the human mind. The finite, manageable number of genes carries the promise that scientific advances eventually will permit humans to know each of these molecular deities intimately, and perhaps even to contemplate reshaping the genes to our own preferred images.

Enigmas Posed by Beneficent Genes

The apparent perfection of the molecular processes underlying life raises a number of questions, both providential and scientific. Challenges to providential interpretation stem from numerous details of molecular operation that give every indication of historical legacy as opposed to *carte blanche* rational design. Challenges to scientific explanation require an account of how natural selection can mold genetic variation at rates and in patterns consistent with the fineness of molecular adaptation.

Providential Enigmas

Assuming that an omnipotent creator wished to deal in the material (as opposed to ethereal) realm,[16] then perhaps there is no special mystery as to why humans and other life should be constructed of physical components such as the carbon, hydrogen, oxygen, and nitrogen atoms so readily available in the elemental world. Neither is there any particular mystery as to why molecular and biochemical pathways then might be necessary to support the physical and chemical operations of life. One would think, however, that a material-delving god with human well-being at heart would ensure that the mechanistic processes operate flawlessly, in accord with the best-conceived design and implementation strategies. Unfortunately, human anatomies and physiologies are subject to abject failure, and even when functioning properly fall far short of designer perfection. At cellular and molecular levels too, close inspection reveals flaws in even the most properly operating of biochemical pathways.

Consider details of the oxidative phosphorylation pathway for ATP generation on the cell's mitochondrial membrane. First, some background about mitochondrial origins is required. Researchers long have suspected some evolutionary link between mitochondria and bacteria. Mitochondria certainly look like bacteria with respect to their small, streamlined, protein-naked genomes, tightly coiled into characteristic circles of DNA. Physiologically, the protein synthetic apparatuses of both mitochondria and bacteria display a sensitivity to several common antibiotics (such as erythromycin, streptomycin, and chloramphenicol), a disposition not shared by the protein-synthesizing machinery employed by a cell's nuclear genes. The most striking similarity between mitochondria and bacteria, however, involves the DNA blueprints themselves. Strings of adenine, thymine, cytosine, and guanine (A's, T's, C's, and G's) in the rRNA genes of mtDNA resemble much more closely those of certain bacteria than they do counterpart genes housed within the nuclear genomes of higher organisms.[17] If a visiting scientific investigator from Mars were shown the relevant nucleotide sequences of human nuclear DNA, bacterial DNA, and human mtDNA, she might conclude that the latter had been mislabeled as coming from a human source.

Scientists here on earth have reached similar conclusions. Based on the astounding similarities between the characteristics of mtDNA and those of purple photosynthetic bacteria, a consensus scientific view has emerged that the present-day mitochondrial

genomes of higher organisms represent the descendants of bacterial
ancestors that much earlier in evolution entered into endosymbiotic[18] relationships with proto-eukaryotic host cells bearing precursors of the nuclear genome.[19] Over evolutionary time, some of
the genes originally carried by the bacterial invaders were transferred to the nucleus,[20] whereas others were retained by the mitochondrion. These remarkable findings give a whole new slant on
human makeup. Literally, our cells are coinhabited by a now
well-integrated amalgamation of genes that originated in completely separate presymbiotic microbes.

Before your skin starts to crawl, understand that these microbial
mergers leading to the first eukaryotic cells took place more than
one billion years ago. Over the long interim, the genomic associations have been much modified and honed by natural selection. In
a sense, we are in part bacterium, but we could no more live
without our mitochondrial associates than they could survive
autonomously without us. "They" and "we" have become inextricably one.

The nature of this mitochondrial-nuclear collaboration is both
a source of wonderment and a striking testimonial to the history-
laden, nonsensical design motifs of molecular associations. Many
biochemical interactions between products of the nuclear and
mitochondrial genomes are intensely intimate. For example,
within the oxidative phosphorylation oval pictured in Figure 4.2
(p. 86) exists a series of molecular protein complexes (respiratory
units I–V, not shown) that govern the flow of hydrogen electrons
and protons necessary to produce most of the 120 watts of power
that energize a man, woman, or child. Each respiratory complex
itself is a joint venture of mitochondrial and nuclear genes. Seven
of the more than twenty-five polypeptides in complex I are specified by mitochondrial genes, the remainder by nuclear genes; three
among the thirteen subunits in complex IV are mtDNA encoded,
and so on. As Doug Wallace explains, the mitochondrial symbiont
"ensures its own survival by keeping its fingers on the jugular vein
of cellular energy flow."[21]

Why in the world would an omnipotent biochemical engineer
jury-rig such a molecular patchwork to perform the most indis-
pensable of metabolic functions? Why should genes of bacterial

origin (mitochondria) be utilized at all to govern critical energy production in the cells of higher animals, humans included? Perhaps there is some (meta)physical necessity for this state of affairs that we don't yet understand, but if so, why wouldn't a creator at least have the good sense to demand that these mitochondria efficiently complete the task of oxidative phosphorylation without the inefficient and cumbersome complications of nuclear gene involvement? How can our metabolic fates be left in such a precarious position? Evolution provides the best explanation.[22]

Besides metabolic pathways, our genomes are also a patchwork of legacy-laden bits of DNA seemingly held together by slapdash mending and darning, not well-considered tailoring. Most structural genes are subdivided into bits and pieces by noncoding intron sequences that must be spliced out of the genetic message before it can be processed properly into a functional protein. Regulatory sequences lying adjacent to structural genes are operationally fragile, all too easily mutated or displaced from their preferred chromosomal precincts with sometimes disastrous consequences to the ill-fated cell. Mobile DNA elements transpose here and there like mischievous rovers, often inserting into and disrupting functional genes. Dead genes (pseudogenes) lie scattered about like functionless corpses in an otherwise bustling cell. Conspicuous junkyards of dispensable DNA, sometimes in long trains of tandemly reiterated units, dot the genome. Tens of thousands of random mutations create never-ending crises, requiring the immediate attention of a cell's emergency crews (repair enzymes and associated apparatus). As if this pandemonium were not enough, during every organismal generation of sexual reproduction, entire chromosomes or portions thereof are wrenched asunder and reassorted during meiosis, then unceremoniously thrown together with chromosomes from another individual during fertilization and ruthlessly tested by natural selection for functionality. Little wonder that the most common fate of a newly formed human zygote or early embryo is death.

In summary, the molecular evidence for evolutionary legacy is ubiquitous, revealing details of molecular operation and genetic descent that challenge theological claims for omnipotent perfection. As was true also for various morphological and other phenotypic features, the molecular machineries of humans and other living organisms give every indication of historical contingency, "evolutionary tinkering,"[23] and *ad hoc* improvisation, as opposed to intelligent design by an attentive supernatural engineer.[24]

Nonetheless, an avowed creationist might well ask, "How can scientists presume to know what would constitute optimal molecular design?" Fair enough! However, the sentiment of this question cuts both ways. If scientists are not allowed to conclude that existing adaptations are less than perfect because of uncertainty about what constitutes molecular perfection, then neither can creationists claim to know and then advance molecular perfection as an argument for rational design. This leaves an intellectual conundrum for assertions about providential perfection. If imperfection truly does exist in the cellular and molecular processes underlying life (as is scientifically insuperable), then it is illogical if not blasphemous to attribute the origin and maintenance of these processes to a god. On the other hand, if the appearance of flaws is illusory only, and the existence of a god assumed, then that deity apparently has gone to great lengths to bamboozle intelligent beings as to his or her intent.[25]

Scientific Enigmas

The scientific view is that life's imperfections exist because genetic and molecular operations are products of evolution, with all of the idiosyncrasies, historical legacies, and contingencies that are inherent in this natural process. Given the sheer evolutionary quirkiness of many genetic adaptations, it is perhaps surprising that molecular machineries work as well as they do. In the light of evolution, the beauty of molecular operation (despite its many blemishes) poses significant explanatory challenges that should not be cavalierly dismissed.

Of course, evolution is not at all the random process that many people assume. Natural selection promotes the very properties that both evolutionary biologists and creationists seek to understand: adaptations. True, the raw materials upon which natural selection operates (mutations, broadly defined) are thought to arise with adaptive haphazardness, but genetic variants that survive the selective filter form a systematic, adaptive subset of the originals. George Williams explains this in response to a critic's objection that organismal features, by analogy, are no more likely to evolve than that a perfect copy of Hamlet's soliloquy should appear from a monkey playing with a typewriter. Williams counters that this randomly typing monkey (or a logistically tractable but equally illiterate computer) *would* in fact reproduce the soliloquy in short order provided that selection preserved all typing that made the computer-monkey's cumulative effort resemble Hamlet's words more closely, and rejected all changes that decreased the resemblance.

Many of the genomic features and processes mentioned above as evidence for imperfection in molecular design no doubt contribute to the genetic fodder for selective tinkering. For example, the introns that punctuate coding regions of structural genes have been hypothesized to facilitate an evolutionary process known as "exon shuffling." The theory is that exons often encode discreet functional domains of proteins, such that genetic recombination among selectively prefabricated modules occasionally may result in the assembly of new mosaic genes with novel metabolic capacities. Even if effective exon shuffling is relatively rare in evolution, the vast time scales involved make adaptive proliferation of functional capabilities plausible by this genetic mechanism. A separate process by which introns are known to promote molecular novelty, in this case within the lifetime of an individual, involves "alternative splicing" of RNA. After intron regions have been cleaved enzymatically from a particular mRNA transcript, the remaining coding segments from separate exon domains sometimes are spliced together in alternative formats, resulting in proteins with varied structures and functions.[26]

Evolutionary tinkering also takes place when regulatory genes mutate or are transposed, thereby introducing metabolic alterations for selective scrutiny. Transposable elements, even those that behave selfishly, occasionally generate mutations beneficial to their hosts. Many of these mutations may involve regulatory changes in the expression of structural genes. Indeed, some believe that regulatory mutations, defined as those that alter the temporal or spatial patterns of gene expression without necessarily altering protein coding sequences, may be the real "stuff" of a great deal of morphological and behavioral evolution.[27]

The common process of gene duplication provides another source of variation for evolutionary fiddling. After a structural or regulatory gene has been duplicated, one copy may continue to perform the original function whereas the second is free to roam in new operational directions. This evolutionary wandering, fueled by mutation and continually scrutinized by natural selection, most often leads nowhere (as evidenced by the ubiquity of pseudogenes). On occasion, however, the duplicated genetic nomad may stumble onto innovative selective pathways leading to novel functional occupations within the cell.[28] Sometimes, even dead genes may be brought back to life.

Several additional evolutionary processes are exemplified nicely by the globin family of genes (see Figure 4.3).[29] Functional hemoglobin molecules that transport oxygen to our tissues are composed of two types of polypeptide subunits, α and β. The α-chain subunits are encoded by three functional genes ($\alpha1$, $\alpha2$, and ζ) clustered on human chromosome 16; β-chain subunits are produced by five operational loci (β, δ, ϵ, Aγ, and Gγ) on chromosome 11. These subunits variously combine to form hemoglobin molecules specialized for particular tasks. For example, the ζ, Aγ, and Gγ chains participate in formation of fetal hemoglobins, whereas α, β, and δ chains are utilized in adult versions of the protein. Nestled among these active loci are three nonfunctional hemoglobin pseudogenes ($\Psi\zeta$ and $\Psi\alpha1$ within the α family, and $\Psi\beta1$ within the β family). The evolutionary diversification of globin genes has been accomplished by recurrent gene duplications, at least one chromosomal transposition, incorporation of numerous mutations in the various gene lineages, and functional differentiation among the family members. Genealogical relationships among these genes and pseudogenes are evidenced by their particular nucleotide sequences and by their chromosomal addresses.

Further evidence for evolutionary homology (shared ancestry) in the globin family is that all hemoglobin genes have three exons (and hence two introns). Exon 2 encodes the heme-binding domain of the protein. More distant evolutionary cousins of the hemoglobins are genes that encode other oxygen-binding mole-

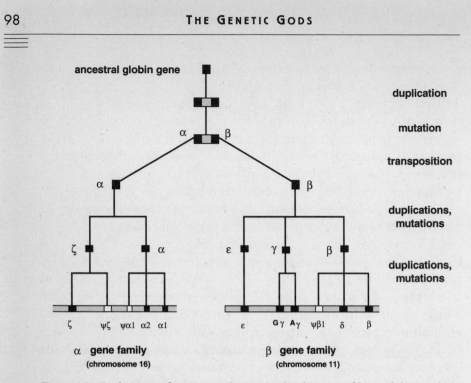

Figure 4.3 Evolutionary factors contributing to the diversity of hemoglobin genes in humans. The sequences $\Psi\zeta$, $\Psi\alpha1$, and $\Psi\beta1$ are pseudogenes.

cules: myoglobins of animals and the leghemoglobins of plants. Myoglobin displays the same tri-partite exon structure as do the hemoglobins, whereas leghemoglobin contains three introns, one of which splits the exon corresponding to the heme-binding domain of the hemoglobins. One intriguing likelihood is that exon 2 of the globin genes was derived by a genetic shuffle (exon fusion in this case) in the ancestral, great-great grandfather of the extended superfamily of globin genes.

The textbooks and journals of molecular biology are filled with case histories of this sort in which multitudinous aspects of evolutionary monkeying appear to have shaped the structures and resulting functions of extant genomes. Yet, can these selectively guided genetic meddlings truly explain the marvelous molecular operations of organisms, or life's great diversity? Biologists long have sought alternative scientific explanations that might account for the elegance of organismal adaptations. For example, one provocative hypothesis elaborated by Stuart Kauffman is that inherent wellsprings of biological order exist outside the framework of conventional Mendelian and Darwinian mechanisms alone.[30] Under Kauffman's speculation, spontaneous processes of molecular self-organization facilitate the action of natural selection in achieving biological order. This hypothesis remains to be tested scientifically.

Somewhat less revolutionary is a suggestion by Chris Wills that genes over the long course of evolution tend to get better at the evolving process itself.[31] In addition to entailing new modes of gene organization and interaction, such "evolutionary facilitation" might be made possible by a genome's evolved capacity to capitalize upon nontraditional (newly discovered) mutational forces as a source of genetic variation for selective scrutiny. For example, one mechanistic possibility worthy of further experimental analysis is that environmental stresses may trigger particular classes of heritable mutations (such as those induced by transposable elements) that might be nonrandom with respect to adaptive utility.[32] Perhaps some transposable elements are stimulated to move in particular adaptive ways in cells when their organismal bearers experience stressful environmental conditions.

A pregenetic version of a "directed mutation" evolutionary hypothesis originally was advanced by Jean-Baptiste Lamarck nearly two hundred years ago.[33] Lamarck suggested that organisms acquire heritable adaptations during their lifetimes through the continual exercise of particular organs or other body parts. By stretching their necks to reach higher branches, giraffes acquire greater neck musculature and length, a disposition transmitted directly to progeny.[34] If true, the inheritance of favorable characteristics acquired during the lifetime of an individual would at once alleviate the enigma of adaptive randomness of mutations under traditional Darwinian evolution, and perhaps explain more readily the perceived levels of perfection in adaptive features such as the giraffe's neck or the exercised human brain.

Indeed, Lamarckian modes of inheritance might be expected to evolve (or at least be selectively favored) under Darwinian evolution. Suppose that a workable molecular mechanism existed for the incorporation of acquired, adaptively relevant genetic information to germline DNA. To the extent that environmental challenges remained similar or predictable across generations, individuals who could transfer genetically such acquired adaptational dispositions to their progeny should display higher genetic fitness than those who could not. Thus in theory, any genetic variation in the propensity for adaptive Lamarckian transmission could be seized upon by Darwinian natural selection, with the net result that genes conferring Lamarckian capabilities would increase in frequency and come to characterize natural populations.

Scientific quests for evidence of Lamarckian inheritance continue, but most such explorations have failed. Hereditary mechanisms as currently understood appear mostly inconsistent with this evolutionary mode.[35] Although a molecular apparatus compatible with Lamarckian inheritance remains to be discovered, Mendelian analogues that confer similar types of adaptational advantage are ubiquitous. These involve genetically-based dispositions for appropriate somatic responses to regularly encountered environmental challenges. Examples include many adaptations[36] that are so common they are taken for granted.

Why do we feel pain? Pain alerts us to body-threatening situations. Why do we tan upon exposure to sunlight? Tanning protects us from UV radiation. Why do we cough, sneeze, and blink? These remove irritants. Why do our bodies sweat, shiver, sleep, heal wounds, pump adrenaline, and build muscles? These functions answer the environmental challenges of temperature, stress, and injury, among others. At the cellular and molecular levels, why are salivary and other digestive enzymes secreted in Pavlovian fashion at the sight or smell of food? Why are the numerous pathways of cellular metabolism regularly switched off and on? Why are molecular mechanisms intermittently activated for the replication, transcription, translation, degradation, and repair of nucleic acids? Why are appropriate antibody responses mounted by the body's immunological defenses? The general answer to these and related questions is that these adaptive capabilities represent genetically-based competences that evolved under the influence of natural selection. In effect, we and other species *have* achieved many of the potential advantages of Lamarckian inheritance, but through the mechanisms of Mendelian heredity. Transmission is not of genes directly altered by environments, but rather of genes that impart metabolic and cellular scopes for appropriate rejoinders to environmental interrogations. In other words, it is the genetic capacity for pertinent phenotypic responsiveness that is inherited. This is one good reason why more attention should be directed toward identifying and fostering the kinds of social and physical environments for humans that suitably complement our evolved response capabilities.

In only a few cases have the particular genes involved in phenotypically plastic responses been pinpointed. One good example involves genes of the immune system. Humans can produce many millions of different antibodies depending on the antigenic (environmental) affronts. Antibodies (or immunoglobulins) are Y-shaped proteins, produced by a special class of white blood cells known as B cells that recognize exclusively the species-idiosyncratic antigenic determinants (epitopes) displayed on the cell surface of an invading microbe or bit of foreign tissue. Immunoglobulins consist of structural subunits encoded by three clusters of antibody genes, located on human chromosomes 2, 14, and 22. One of the great accomplishments of molecular biology has been understanding how an astronomical diversity of antibodies is generated by these genes. The process is complicated, but includes extensive recombinational shuffling within the gene clusters during somatic B cell maturation, and varied amalgamations of the resulting protein subunits during antibody formation. The net result of these combinatorial aspects of antibody assembly is that perhaps 100 million different types of antibodies may be produced by the collective population of B cells in each individual. Some of these antibodies fortuitously may produce a molecular match to the epitope of a particular microbial invader. The relevant B cells then proliferate clonally as the body mounts an immune response that (hopefully) quells the microbial invasion and makes us feel better.[37]

The immunological system provides a wonderful example of how genes responsible for phenotypic responsiveness to environmental challenges can be transmitted in Mendelian fashion without themselves being heritably altered by the environment. In this case, the mode of biotic response also happens to make a great deal of engineering sense. The potential pool of microbial disease vectors and hence the numbers of different antibodies necessary to mount an effective defense are vastly larger than the total number of genes within the human genome, thus precluding a one-gene-one-antibody response mode. The evolutionary solution has been the elaboration of a heritable, gene-based capacity for appropriate somatic differentiation in response to unpredictable disease challenges.

Organisms cannot be assumed to possess genetic machineries to withstand environmental affronts to which their ancestors have not been exposed consistently. However, for certain types of disease challenges, genomes have evolved coping devices that often confer upon somatic cells the regulatory or developmental flexibility to accommodate the challenge of the moment, be it the healing of a wound, the digestion of a meal, or the destruction of a disease microbe. It remains to be determined whether the biological products of Darwinian evolution have capabilities that extend further— for example, whether genomes exert any influence over the course of their own evolution by generating particular types of mutations (or at particular rates) when needed. Thus far, the scope of genomic responsiveness has been sufficient to accommodate the evolutionary continuance of all extant lineages. Let us hope that the novel global environments currently being fostered by human activities will not fall outside the compass of appropriate response by our own genomes or those of the planet's other biota.

Apart from molecular considerations, we can see rather directly how selection can account for organismal adaptations and the diversity of life through our manipulation of the selection process. Ever since humans began domesticating animals and plant crops some 10,000 years ago, and certainly ever since Darwin,[38] humans have mediated natural selection, promoting extremely rapid evolution in a variety of organismal traits. Kernal production in maize, acridity in peppers, milk yield in cows, docility in sheep, galloping speed in thoroughbred horses, plumage coloration in parakeets (all wild budgerigars in Australia are forest-green), and color and corpulence in goldfish and koi are but a few of the thousands of traits that have been artificially selected to an astonishing degree and at a remarkable pace by animal and plant breeders. Whatever the particulars of the molecular genetic alterations, the unassailable empirical reality is that nearly any organismal feature can be evolutionarily altered by artificial selection, no matter how arbitrary or otherwise absurd to the organism (or to the observer) that evolved feature may be. A decorative goldfish may be disfigured and debilitated by its gargoyled body with absurdly bulged eyes and hopelessly flowered fins, but so long as humans prize these features and promote their retention and elaboration through selective breeding and careful nurturing, they are made to appear. Can we expect less from natural selection in promoting adaptations to nature's demands?

To introduce the concept of organismal responsiveness to selective influence, consider the peppers (genus *Capsicum* in the nightshade family Solanaceae). From the fiery Tabasco and jalapeno peppers to the sweet pimientos and giant bells, from rotund red and yellow waxy ornamentals to lantern-shaped habaneros to long slim cayennes and green chili peppers, a veritable cornucopia has been produced by selective farming.[39] Indian cultivator-breeders began the domestication of peppers about four thousand years ago, utilizing five of the twenty to thirty wild species of *Capsicum* native to Central and South America. Today, approximately sixteen hundred varieties of domesticated peppers are recognized. Peppers were unknown outside the New World until Christopher Columbus introduced them into Spain in 1493. Cultivation soon spread throughout Europe and Asia, and today many cuisines such as Indian, Thai, and Szechwan would hardly be the same without these tantalizing New World condiments.

Another of my favorite examples of artificial selection involves our most familiar friend, the domesticated wolf *Canis familiaris*. In a scant ten millennia of selective breeding,[40] a spectacular diversity of dog varieties has emerged, ranging from twenty-four-ounce chihuahuas to two-hundred-pound Saint Bernards. Nearly every physical feature has proved selectable, from heads to tails.

Dispositions and behaviors of canines have proved to be selectively moldable as well, an observation not without relevance to broader discussions of genetics and ethology. Among the sporting or working breeds of dogs are evidenced behavioral proclivities for: hunting down large quarry (great danes and rottweilers originally were bred to hunt wild boar, borzoi to hunt wolves, salukis to chase gazelles, and Rhodesian ridgebacks to challenge lions); the dogged pursuit of small or burrowing prey (by dachshunds and many terriers); pointing of game (by pointers and setters); flushing (spaniels); retrieving (retrievers); fetching in water (newfoundlands, Irish water spaniels); baying (beagles); tracking by olfaction (bloodhounds); herding (collies, sheepdogs, and schnauzers); hauling (huskies); racing (whippets, greyhounds); and fighting (mastiffs, boxers, and some terriers). From high-strung Italian greyhounds to laid-back English bulldogs, from amiably stubborn otterhounds to obedient King Charles spaniels, from loud-voiced dachshunds to nonbarking basenjis, and from aloof dobermans to cuddly toy dogs such as the bichon frise, an impressive degree of behavioral as well as morphological canine diversity has evolved in a short time under the selective demands of humans. If these diverse creatures were known only as ancient fossil remains, they might have been considered separate species, or placed into distinct taxonomic genera or families.

Plants and animals also can evolve rapidly in response to natural pressures in their environment. The Hawaiian silversword alliance consists of twenty-eight plant species, all descended from a single ancestor that colonized the islands within the last few million years. These species have radiated adaptively to fill a wide diversity of niches on the Hawaiian Islands, and include forms as diverse as vines, trees, shrubs, rosette plants, and cushion plants—a tremendous morphological diversity that belies a close ancestry revealed in their molecular genetic makeup.[41] In the animal world, all living cats (thirty-eight species in some eighteen genera) have arisen and diversified within about the last 10 million years,[42] adaptively radiating into forms as distinct as the diminutive ocelots and margays to monstrous lions, tigers, and jaguars. The domestic cat itself has achieved a considerable diversity of forms under the far more recent influence of human selective breeding.

Most organismal genomes have been so responsive to artificial and natural selection, the puzzle is not how the earth's exuberant biotic diversity could have evolved over geological timescales, but why some evolutionary lineages and taxonomic groups remain morphologically static over long periods of time. All four species of horseshoe crabs alive today are nearly identical in external morphology to horseshoe crabs preserved as fossils 150 million years ago.[43] For such groups, evolutionary processes confined long-term morphological differentiation within boundaries that are far narrower than they theoretically could be given the rapid pace of short-term phenotypic change commonly observed in most species under strong directional or diversifying selection. One possible explanation is that the relevant selective regimes, such as the environment or predators, have remained stable for long periods of time. Another possibility is that morphological change has been limited by phylogenetic or ontogenetic constraints. Even in rapidly evolving groups, historical contingencies clearly limit the range of evolutionary outcomes to a small subset of what might be envisioned by Dr. Seuss.

FIVE

Strategies of the Genes

Heredity

I am the family face;
Flesh perishes, I live on,
Projecting trait and trace
Through time to times anon,
And leaping from place to place
Over Oblivion.

The years-heired feature that can
In curve and voice and eye
Despise the human span
Of durance—that is I;
The eternal thing in man,
That heeds no call to die.

Thomas Hardy, 1917

The ancient Greeks had many gods, each with particular strengths, weaknesses, and operational jurisdictions.[1] The Greek gods collectively displayed a plethora of foibles as well as virtues. They could be jealous, lustful, spiteful, angry, moody, hateful, and selfish, but they also could be loving, stoic, brave, nurturing, generous, and altruistic. The genetic gods, in all their multiplicity and eccentricity, resemble the Greek gods more closely than they resemble the God of Christianity or the Allah of Islam. Like the Greek gods, each operational gene has a primary functional jurisdiction. Furthermore, genes can be magnanimous in providing for their human hosts, but they also can be hurtful and uncaring. In their interactions with one another, genes often are collaborative and mutually supportive, but they also can appear divisive and self-serving. These multiple personalities reflect the hereditary struggles of the genes to reproduce and avoid death.

Fundamentally, these alternative faces of the genetic gods have come about because separate genes have quasi-independent evolutionary fates under sexual reproduction. Owing to the gene-shuffling processes of meiosis and fertilization, the hereditary pathways of unlinked genes through the germ lines of an organismal pedigree are not perfectly coincident. Under the rules of Mendelian heredity, particular gametes transmit partially randomized subsets of parental genes to progeny. Thus, in the never ending scramble for replicative continuance, each successful gene in effect devises, under the tutelage of natural selection, a quasi-personalized evolutionary strategy that tends to enhance its prospects for successful transmission from one generation to the next. Fortunately for us, contributing to the well-being of the individual is an effective gene strategy, because this usually enhances reproductive success of the organism and thereby improves the likelihood that the gene's descendants will be represented in subsequent generations. However, other strategies also exist that can profit a gene at organismal expense, producing conflicts of interest between the renegade gene and its more civic-minded subcellular compatriots. In the final analysis, all genes are reproductively selfish, but the routes to success are varied.

In biology, organisms often are viewed as complex machines, with genes providing the coded instructions for proper manufacture and assembly of the component nuts and bolts. A more useful metaphor might view genes as members of intraorganismal social groups that display elaborate divisions of labor and interdependencies. Particular genes help to shape such subcellular societies, but they also are subject to constraints imposed by these societies on a gene's individual freedoms. When organismal genomes are described metaphorically as social collectives of interacting DNA sequences, it is easier to imagine how disputes, cooperative behaviors, and social contracts might arise, and how they might shape the evolutionary strategies of genes.

Suppose that each genome was transmitted intact, such that

offspring were identical genetically to the parent (barring mutation). The evolutionary fates of all genes within the lineage would
be coupled, and whatever was in the best reproductive interest of
any one gene would coincide with that of the many. Under such
asexual reproduction, no evolutionary incentive would exist for
particular genes to adopt agendas that depart from the collective
genomic good. Because the genome would survive or fail as a unit,
natural selection at the levels of both the gene and the individual
would tend to favor a "one gene for all and all for one" mentality
for the DNA sequences. Many microbes, plants, and invertebrate
animals commonly reproduce asexually, as do a handful of parthenogenetic fishes, reptiles, and other vertebrates.[2] Any negative
intracellular interactions among the genes of strict asexual reproducers should reflect harmful mutations not yet eliminated by
natural selection,[3] rather than any overtly selfish motives of individual genes at the expense of others.

In humans, however, as in most other higher animals, genes are
sorted and recombined during every generation of sexual reproduction. As each individual human begins to age, genes in the germ
line scramble onto gametic lifeboats (eggs and sperm) that offer
their only hope for physical survival and continuance beyond the
current generation. On any such lifeboat, each gene will find itself
in the company of a partially randomized complement of the ship's
original crew: a motley assortment that nonetheless includes (under
Mendelian rules) representatives from each of the individual's
chromosomal departments. If the egg or sperm survives, it may
encounter and collaborate with a gamete from the opposite gender.

Imagine yourself as a gene on a gametic lifeboat faced with the
problem of survival. Certainly you would hope to find yourself in
the company of congenial and competent genes, a partially randomized subset of your original shipmates. Even one rotten gene
could spell disaster. If you could earmark lifeboat seats specifically
for your use, that would be desirable from a selfish perspective. If
you were able to replicate yourself for placement onto additional
lifeboats, that option too would enhance the probability that some
of your clonemates would survive at sea. On the lifeboat itself, you
would wish to be useful, or at least not get in the way. If there
were subversive elements onboard, you might wish to silence or

ameliorate their negative influences to the extent possible. Perhaps your skills as a gene are suited best for construction or operation of a new individual, so when your gametic lifeboat joined with another you would set to work for your own and the collective good. If some of your genetic shipmates consistently showed up together generation after generation in the cellular armada, there might be a strong temptation for the lot of you to form close alliances over time, and to thwart any derelicts, mutineers, or pirates who might contemplate a takeover.

All of these strategies and others potentially are available to genes under sexual reproduction. Of course, genes do not consciously weigh the alternatives and make cognizant choices about how to achieve successful transmission to subsequent generations. Nonetheless, the genes that survive will be those that have displayed behaviors enabling them and their descendants to withstand the incredible rigors of eons of natural selection on the turbulent seas of evolution.

Perhaps least intriguing from the perspective of evolutionary strategies are the many genes that are good citizens. These include the beneficent genes, whose proper functions are critical to the survival and reproduction of the individual. No conflict of interest exists between what is best for the gene and what is best for the individual, nor is there any overt conflict of interest among these genes themselves. In quintessential form, each such gene exists as a single copy at one chromosomal location in the genome and carries out a specified functional or structural role that contributes to the collective genomic effort. Many genes involved in development and basic cellular metabolism qualify, as do their regulators and modulators. Functional collaborations among such genes and among their protein products can be breathtakingly refined. Of course, beneficent genes can mutate undesirably, but both the genes and their human bearers typically suffer in these molecular malfunctions. Thus, such mutants tend not to survive for long.

Given sexual reproduction and the attendant physical shuffling of DNA each organismal generation, what evolutionary strategies should work best for beneficent genes? Perhaps the primary challenge is that a gene's proper function must be maintained across a great variety of genetic backgrounds that differ unpredictably between generations. A capacity to work well with other genes in unpredictable combinations is one quality that enhances a gene's probability of survival.

One of the long-standing paradoxes of evolutionary genetics is why genes that collaborate well haven't amalgamated themselves more often into multigene complexes predictably co-transmitted through gametes into the next generation. At face value, such a strategy would seem to enhance genetic fitness by cementing any evolved intergenic partnerships. Yet in most cases, the different genes of a functional complex are assorted independently and recombined during sexual reproduction.[4] If natural selection is in charge, why hasn't the genome (or at least large functionally integrated portions thereof) congealed over evolutionary time into co-inherited units? Why aren't successful genetic teams, once assembled, held together?

This deceptively simple question lies at the heart of the profound issue of why sexual reproduction itself exists. The topic is elaborated later, but one important element for present consideration involves "Muller's ratchet," a phenomenon named after a well-known evolutionary biologist of the mid-twentieth century. Hermann Muller pointed out that in an asexual population of finite size, no individual in the population is likely to be free of deleterious mutations in all genes. Thus, an asexual population over multiple generations can only remain the same or ratchet upward in terms of genetic load.[5] This is because there are only two fates for a mutant clone. Either it goes extinct, and the population's genetic load returns to the former state, or it increases in frequency, and the population's genetic load is higher than before. By contrast, a sexual population may continue to regenerate some nonburdened genotypes owing to the gene-shuffling processes of molecular recombination. This idea, which can be mathematically framed,[6] reminds us that asexuality (the absence of genetic recombination) can cement disadvantageous combinations of genes as well as advantageous ones, indeed perhaps with greater certainty since most mutations injure rather than aid the organism. Muller's ratchet identifies one of several plausible fitness benefits for recombinational processes that are the hallmark of sexual reproduction.

Under sexual reproduction, conflicts of interest among inhabitants of the intracellular community of genes arise whenever the reproductive interests of certain genes come into opposition with those of others. The stage for such feuds commonly is set, for example, by inherent asymmetries of the hereditary process that can lead to overt confrontations whenever particular genes adopt reproductive strategies that selfishly enhance their own fitness to the detriment of other genes or of the organism. Natural selection is the final mediator in all such disputes, but the settlements typically involve compromises on the part of all the participating genes. They also involve tradeoffs between the various hierarchical levels (genes, individuals, families, and sometimes larger groups) at which natural selection may operate. Some of the more evident provocateurs of gene-gene disharmony and evolutionary bickering are discussed next.

Meiotic Drivers

Autosomal Genes

Textbooks of introductory genetics tend to portray meiosis as scrupulously fair under Mendel's law of segregation. In diploid heterozygotes, the two alleles of each autosomal gene have statistically equal probabilities of passage through gametes to the next generation. But what if alleles occasionally arise by mutation that can bias the meiotic process in their favor? By so bettering their transmission odds, such "meiotic drive"[7] alleles would be strongly favored by natural selection at the genic level (see Figure 5.1). If unopposed by counterbalancing forms of natural selection, these meiotic drivers will tend to evolve to fixation (100 percent frequency) in any population in which they arise.

Meiotic drive is not merely a hypothetical possibility; several examples in species other than humans are well known. Some fruit flies, for example, carry a segregation distortion (SD) allele on the second chromosome that causes males who are heterozygous (have one copy of the SD allele) to transmit that allele to more than 95 percent of their progeny. In many house mice populations, t alleles on chromosome 17 are driven to rather high frequency (e.g., 10–20 percent) despite the fact that t/t males who are homozygous

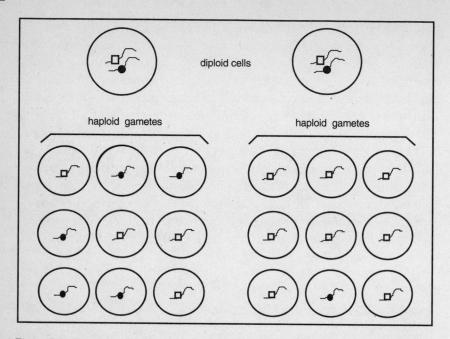

Figure 5.1 Transmission advantages to an allele of meiotic drive behavior. On the left is the normal Mendelian situation in which the two alleles (square and closed circle) of a diploid heterozygote have approximately equal likelihoods of being transmitted to progeny through the gametes produced by that individual. On the right is depicted a meiotic drive allele (square) that can influence meiosis or gametic survival in such a way as to increase its representation in the pool of gametes.

(carry two copies of the *t* allele) are sterile and have reduced viability. This increase in frequency occurs because males heterozygous at this locus transmit mostly *t* alleles to their progeny. In general, natural selection tends to bless even those meiotic drivers that have moderately harmful effects at the organismal level.

Given the transmission advantages that accrue to meiotic drive alleles, why don't they occur more often? In other words, why is meiosis normally so equitable at most loci? Two conventional explanations exist: first, that only a few genes are capable mechanistically of influencing transmission likelihoods via meiotic drive;[8] and last, that irresponsible driving behavior tends to be policed by unlinked genes. This point warrants elaboration. As in the *t*-allele case, genes that distort segregation ratios in their favor usually affect the transmission of other genes adversely, and hence come into strategic conflict with them. Sex-linked meiotic drivers, which will be discussed in detail later in this chapter, offer particularly clear examples. Thus, when segregation distorters arise, so too do selection pressures favoring modifier genes elsewhere in the genome that suppress or override the segregation-distortion behavior. Furthermore, genes whose personal fitness is harmed by meiotic drive normally outnumber any linked genes that profit from the selfish actions of a meiotic chauffeur. Thus, under the collective, selection-mediated legislation of the broader "parliament of genes,"[9] genetic modifiers tend to evolve that in effect revoke the licenses of the meiotic drivers and get them off the hereditary road.

Y-linked Genes

Because the Y chromosome is transmitted unisexually (through males), it provides an excellent example of how a gender-based asymmetry in genetic transmission can lead to contention among genes and to selection pressures for conflict resolution. Suppose you were a self-interested gene located on the Y chromosome. Your prospects for representation in the next generation would be enhanced if somehow you could promote the production and dissemination of your own Y-carrying gametes at the expense of other gametic types. This "Y-linked drive" would be to your short-term advantage, but also would bring you into immediate conflict with the preferred strategies of genes housed on the X chromosome (one third of which in any population are passed through males). Your strategy also would be received poorly by nondriving Y chromosomes, by mitochondrial genes whose inheritance is matrilineal, and by X-linked and autosomal genes that have a vested interest in seeing the continued production of both male and female genders through which they are transmitted.[10] Indeed, an extreme form of Y-linked drive ultimately could prove suicidal because continued production of nothing but males inevitably would result in species' extinction.[11]

Nonetheless, captivated by the prospects of short-term transmission success, and blinded to the ramifications of your actions, you might choose to forge ahead with a strategy of Y-linked drive. If you didn't, some other ultra-selfish Y-chromosome alleles in the population surely would. Given the short-term selective advantage of Y-linked drivers over their nondriving counterparts, why aren't they observed more often in humans and other species? Perhaps Y-driving alleles are observed infrequently in extant species because the populations in which they arose have been meiotically driven to extinction; perhaps they are mutationally rare (see note 8); or, perhaps countervening selection pressures on other chromosomal genes have thwarted their spread.

William Hamilton raised a hypothesis that combines elements of these latter two explanations, and also might account for the long-standing enigma of why the Y chromosome alone carries so few functional genes. According to Hamilton's theory, the relative genetic inertness of the Y chromosome is due to the suppressive effects of autosomal (or other non-Y) modifier genes that have evolved under the influence of natural selection to stymie Y-linked meiotic drive behavior. Thus, evolutionarily, the Y chromosome has been stripped of most operationality beyond that minimally required to initiate the developmental cascade of maleness. Although this hypothesis remains unproved, it exemplifies how, in principle, natural selection might resolve internecine genomic conflicts and thereby curb the strategic maneuverings of any unduly self-interested genetic drivers on the Y.[12]

X-linked Genes

Meiotic drive alleles on the X chromosome could occasion similar long-term catastrophes for populations via production of gross excesses of females. In theory, however, such disasters would come about more slowly because the X-driving alleles are somewhat less exposed to natural selection[13] and hence would increase in frequency in the population less quickly. Several cases are known of X-linked meiotic drive (again in species other than humans). In one well-studied case involving fruit flies, X chromosomes with sex-ratio (SR) alleles distort gender ratios in the progeny of males by causing most Y-bearing sperm to degenerate upon completion of meiosis. Thus, males carrying an SR X chromosome typically produce more than 90 percent daughters.[14] These systems also demonstrate how countervening selection pressures have led to the evolution of autosomal modifiers that put the brakes on X-drive behavior: In some of the species possessing SR X chromosomes, suppressor genes have been identified that moderate the distortion of segregation by the SR locus.

Cytoplasmic Genomes

Apart from mitochondrial DNA, cytoplasmic genomes (those found outside the nuclei in eukaryotic cells) also include chloroplast DNA in plants, and various intracellular microbes who sometimes hitch evolutionary rides through germline cytoplasms. Such intracellular microbial parasites are common in insects, for example, and include bacterial spiroplasmas and streptococci that are especially well known in fruit flies. Normally, cytoplasmic genomes are transmitted through females rather than males.[15] Because cytoplasmic and nuclear genomes abide by different hereditary rules, their reproductive fates are uncoupled partially. Thus, subcellular disputes again can arise over preferred evolutionary strategies.

As was the case for sex-linked genes in the nucleus, this gender-based asymmetry of inheritance also generates conflicts of interest with nuclear autosomal genes. Because of their maternal transmission, cytoplasmic genes have no vested interest in residing in males. As a consequence, any cytoplasmic gene mutations whose differential effects on organismal fitness are confined to males should themselves be largely invisible to natural selection (apart from second-order influences operating through properties in the population such as sex ratio or mate availability). Thus, in contrast to mutations at autosomal nuclear loci, cytoplasmic mutations that enhance male fitness show little directive tendency to increase in population frequency because their beneficial effects on males are not rewarded by increased representation among progeny. Similarly, female-transmitted alleles with male-limited deleterious effects may be retained in populations simply because of relaxed selection pressure against their immediate loss.[16] Furthermore, any cytoplasmic allele that biases females toward production of daughters would show a fitness profit (at least in the short term) by increased representation in subsequent generations. In the extreme, such driving behavior by cytoplasmic genomes could result in species extinction (or, perhaps, the evolution of parthenogenesis) through continued production of females alone. Again, these selectively favored tendencies from the perspective of selfish cytoplasmic genomes can expect legislative opposition from the parliament of nuclear genes whose fitness is diminished by these cytoplasmic outlaws.

Many situations are known in which cytoplasmic elements are implicated in reducing male fitness. In insects, cytoplasmic "son-killer" factors are common and can lead to skewed sex ratios. A dramatic example in humans involves a cytoplasmic factor in eggs that reportedly is lethal to Y-carrying sperm.[17] Another mitochondrial mutation produces an early-onset marrow/pancreas syndrome in men, but only lesser, late-onset symptoms in women. In another example, approximately 85 percent of individuals affected by the serious eye disorder LHON (caused by a mitochondrial mutation) are males. Interesting lines of future research will address whether male-female differences in other degenerative conditions, such as heart disease, have a partial basis in cytoplasmic genes.

Cytoplasmic effects in males often are registered as fertility problems. Sperm in higher animals derive their vigor and motility from mitochondrial batteries packed at the base of a propelling tail or flagellum, and both *de novo* and inherited mtDNA mutations in humans are suspected to sabotage this energy source. Male sterility is common in insects also, and frequently is attributable to mtDNA mutations or to cytoplasmically housed bacteria. In plants, the phenomenon of cytoplasmic male sterility (CMS) is widespread, observed in more than a hundred species. As one might predict, the CMS effects in many plant species appear to have been ameliorated by the evolution of contravening nuclear suppressor alleles that restore male fertility. At any point in time, the outcome of nucleo-cytoplasmic strife with the cell is likely to reflect the particular cytonuclear genetic associations contingently evolved under competing selective influences.

Because eggs, rather than sperm, contribute the vast majority of cytoplasm to a zygote, cytoplasmic genes are inherited for the most part maternally. From an evolutionary perspective, a more interesting but unresolved question is whether this uniparental inheritance is merely an unselected consequence of asymmetry in gametic size, or whether it has been selected for specifically (for example, to suppress the more rapid spread of selfish cytoplasmic alleles that otherwise might attend biparental mtDNA inheritance).[18] Other open questions concerning the coevolutionary games between nuclear and mitochondrial genomes include: (a) Is the partial surrender of mitochondrial function to the nucleus (subsequent to the endosymbiotic origin of mtDNA more than a billion years ago) the result of nuclear hegemony, or nuclear infiltration by mitochondrial genes "wishing" to achieve the advantages inherent in recombination, or as some combination of these factors? (b) Why hasn't the surrender (or invasion) of mitochondrial genes to the nucleus gone to completion? and (c) Why should mitochondrial genes remain quasi-autonomous instead of integrating themselves into the nuclear genome?

Finally, there is another salient difference between the transmission modes of cytoplasmic versus nuclear genes. Most nuclear genes in a germ cell lineage pass through a bottleneck of one molecule per gamete per generation, whereas cytoplasmic genomes are transmitted in many copies. This has ramifications for mtDNA molecules competing for transmission across cell generations, and for the replicative strategies of nuclear versus mitochondrial genes in their quests for molecular immortality.

Mobile Elements

Gypsy, tourist, stowaway, castaway, wanderer, pioneer, vagabond, pogo, Magellan, hopscotch, gaijin (Japanese for foreigner), hobo, jockey, mariner—all of these names have been assigned to various members of an astounding array of recently discovered "jumping genes" that appear to be nearly ubiquitous in the genomes of higher organisms. First discovered by Barbara McClintock in her studies of maize in the 1940s (for which she received a Nobel Prize in 1983), mobile elements comprise a broad class[19] of miniature genes that have evolved the capacity to frolic about the genomes of a host organism in elaborate fertility dances initially choreo-

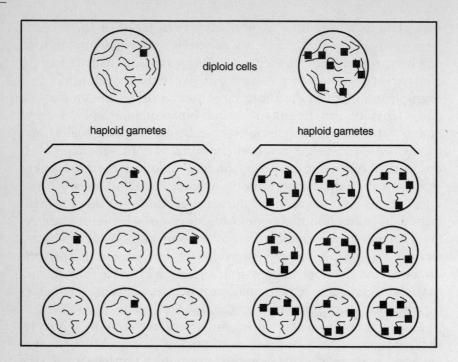

Figure 5.2 Transmission advantages to a dispersed mobile element. On the left is the normal Mendelian situation in which an allele (filled square) of a single-copy gene in a diploid heterozygote is represented in about one-half of the gametes produced by that individual. On the right is a jumping gene that has replicated and dispersed itself to multiple positions across the six pairs of homologous chromosomes shown. Through such behavior, the mobile element has increased greatly the likelihood of its representation in the pool of gametes.

graphed to win favor for their own transmission during sexual reproduction. In effect, like the meiotic drivers discussed above, many of these genetic elements overcome the equal-segregation rules of Mendelian meiosis, but in this case they do so by replicating and dispersing across multiple chromosomal sites. Through such behavior, mobile elements increase their likelihood of survival to subsequent generations through germline cells (see Figure 5.2). Thus, these elements can spread rapidly via the infection of new sexual lineages in the host population. This brazenly selfish behavior has earned these genetic vagrants the reputation of parasites, and with good reason.[20] The fitness advantages to the mobile element often come at organismal expense.

There are several reasons to think that the activities of mobile elements diminish the short-term fitness of their organismal host and its other genetic citizenry. First, mobile elements must place some metabolic burden on host cellular function and genomic replication. Mobile elements occur in astounding numbers in eukaryotic cells, and indeed more than 90 percent of nuclear DNA in some species is composed of jumping-genes or their less frisky evolutionary descendants. Second, mobile elements appear to be a major evolutionary agent of spontaneous (and frequently deleterious) gene mutation.[21] Mobile elements commonly insert into the coding regions of functional genes and can disrupt their normal function. They also generate other classes of mutations that can damage the host. These damages may result when a mobile element inserts into a regulatory gene region, or when it fosters chromosomal instabilities such as translocations, inversions, deletions, and duplications.

Finally, theory suggests that transposable elements will tend to increase from initial low frequencies even in the face of severe counter-selection at the organismal level. Consider a cross between individuals with and without a mobile element. The element will be inherited by 50–100 percent of offspring depending on how efficiently the element replicates to different chromosomal locations. This implies that a new mobile element can spread in an organismal population despite reducing the fitness of individuals by as much as one half![22] Such a genetic load might even become too great for some populations to bear, and could lead to species extinction (although, of course, that would also spell the end of the parasite).

The behavioral similarities of some mobile elements to parasites can be explained by their comparison to retroviruses (RVs). Retroviruses are small single-stranded RNA viruses, found mostly in mammals, that are able to encase themselves in a protective envelope and infectiously transport across the cells of the same or different host individuals. Some retroviruses carry oncogenes (cancer-causing factors) or otherwise harm their hosts. Structurally, RVs share several features with a class of mobile sequences known as retrotransposable elements (RTEs), including possession of a reverse transcriptase gene involved in the transposition process. This raises an interesting evolutionary question. Might RTEs represent degenerate retroviruses that lost the ability to move themselves among cells? Alternatively, RVs may have evolved from ancestral RTEs by acquiring these capabilities. Recent phylogenetic studies of the structural molecular features of reverse transcriptase genes suggest that RVs probably evolved from RTEs.[23] Regardless of their origins, the structural and behavioral similarities between certain mobile elements and viruses are inescapable.[24]

As in any host/parasite association, the host species need not remain a passive observer of mobile element activities, but may participate. Just as exogenous parasites and their hosts often evolve mutualistic compromises, so too may endogenous mobile elements and their landlords work out reasonably amicable solutions to their conflicting priorities. Selection acting on the nonmobile genes should ameliorate or counter any deleterious consequences of the mobile elements, and selection acting on the mobile elements should lead to a degree of self-policing of activity likely to harm the host. The latter is particularly true when a mobile element reaches high copy number in the genome because the marginal fitness gains to be achieved by selfish replication then diminish greatly.

Other theoretical predictions about host–element coevolution are that mobile elements should be more active in germline than in somatic cells[25] (because only there does the presence of dispersed copies enhance the element's transmission probabilities); and show greater activity in naive hosts (because a naive host may not yet have evolved suppressive mechanisms to thwart the behavior of the element). Both predictions have received empirical support.[26]

The fitness benefits that accrue to mobile elements through replicative transposition depend critically on sexual reproduction by the host.[27] As such, mobile elements can be viewed as sexually transmitted diseases. To understand why this is so, consider the fate of a proliferating genetic element strictly confined to an *asexual* lineage. No matter how many copies of the element exist within the genome, the element either will or will not be represented in subsequent generations depending on the fate of the clonal lineage.

Multiplicity within an asexual genome does nothing to enhance a mobile element's probability of successful transmission. By contrast, under *sexual* reproduction, a mobile element that occupies multiple chromosomal sites in the germ line has increased its odds of transmission greatly.

For this reason, a strong evolutionary correlation might be anticipated between the degree of sexual reproduction displayed by a host species and (after factoring out complications of historical legacy) the extent to which repetitive mobile elements are present within the host's genome. Empirically, such expectations generally are met both among and within species.[28] For example, asexually transmitted mitochondrial and chloroplast DNAs of higher organisms essentially are free of mobile elements, whereas sexually recombining nuclear genomes of these same species are riddled with them. Also, the numbers of jumping genes in sexual eukaryotes greatly exceed the tallies of these elements in largely asexual prokaryotes.

Although first discovered in corn and studied most thoroughly in these plants and in fruit flies, these gamboling genes also can be found in mammals, including humans. Despite the fact that organized searches for jumping genes in humans barely have begun, many transposable elements already have been discovered, including a LINE-1 family of sequences that exists in about 50,000–100,000 copies per cell and by itself composes an astounding 5 percent of the human genome! Another mobile element, a 300 bp *Alu* repeat, is represented in more than 600,000 copies scattered about our chromosomes.[29] Among the varied disorders caused by mutations produced by mobile elements are cases of hemophilia, acholinesterasemia, neurofibromatosis, and a significant fraction of instances of lipoprotein lipase deficiency.[30] With further research, many more spontaneous deleterious mutations in humans undoubtedly will prove to be caused by mobile elements.

After the audience at a recent symposium had shuddered at the ubiquity and disasterous effects of mobile elements in humans and other species, a concluding speaker attempted to reassure the listeners that these little intra-genomic nomads really are friendlier than they appear.[31] He was referring to evolutionary speculations and some recent evidence that host organisms also can benefit from their associations with jumping genes. There are at least two general aspects to such weal. First, some of the host benefits probably reflect incidental but fortuitous byproducts of the normal activities of mobile elements. As major mutagenic agents, mobile elements contribute to the necessary pool of genetic variation upon which the long-term continuance of a species depends. In this important sense, jumping genes may be critical for their host populations. However, because of natural selection's lack of foresight, it is doubtful that such mutagenicity evolved explicitly for this purpose.

Second, some host benefits surely represent the selectively driven outcomes of coevolution between mobile elements and the genome. Close company between mobile elements and the host genome may have evolved into a symbiotic relationship not too unlike that between mitochondrial DNA and nuclear DNA. A remarkable fact is that nearly every DNA sequence critical for the regulated expression of eukaryotic genes has been found in one or more mobile elements, suggesting that the latter upon appropriate genomic insertion can influence patterns of eukaryotic gene regulation, and hence evolution. Occasionally, host genomes then might capitalize upon the regulatory potential latent in mobile elements and put them to their own regulatory use. Having once been captured into host service, the element itself also could profit from the developing evolutionary alliance.

A case in point involves the human *Amy1* gene in the amylase family of loci. *Amy1* is active in salivary glands and its enzymatic product initiates starch digestion in our mouths. Detailed molecular analyses have shown that this tissue-specific expression is governed by a cryptic retroviral-like element that apparently inserted near the *Amy1* gene some 45 million years ago. Other documented examples of captured regulatory control involve two human globin genes (α and ϵ) that utilize truncated sequences of the formerly mobile *Alu* elements to regulate expression in a tissue-specific manner. These examples notwithstanding, the extent to which host genomes have confiscated the regulatory potential in mobile elements may be grossly underappreciated, because the footprints of these deeds tend to have short evolutionary half-lives, making them difficult to uncover in the detective work of molecular biology.

In other cases, it is less clear whether the behaviors of jumping genes reflect the outcomes of natural selection on the mobile elements themselves, on host genomes, on interactions between elements and hosts, or none of these. An intriguing recent observation is that significant increases in the transpositional (and mutagenic) activities of mobile elements are associated with host inbreeding, interspecific hybridization, and exposure to stressful environmental factors. Presumably, these are precisely the times at which response mutations are needed most by the host organism, but they also are occasions when mobile elements might most wish to abandon a stressed host genome. Further research on mobile elements under stress should reveal the extent to which transpositional activity is regulated by the host genome as opposed to the elements themselves, the precise nature of the mutations induced, and whether these mutations display any directionality with respect to the adaptive needs of the host organism.[32]

I have emphasized that not all is politeness and pleasantry in subcellular genetic societies. The delicate mix of continuity yet temporal variety in molecular associations under sexual reproduction has resulted in a diversity of ways in which individual genes have responded to the pressures of natural selection to enhance replicative fitness. Many genes behave quite civilly within their cellular communities, but they do so with selfish replicative motives. Other genes interact with antipathies ranging from mild jostling and elbowing within their cramped lineage quarters, to all-out fisticuffs over reproductive turf.

There is no need to ask why a god would countenance such cacophonous elements in his or her molecular symphony. One need look no further than the mechanistic operation of natural selection in the evolutionary context of sexual reproduction. That, however, begs a more general question: Why has sexual reproduction itself come to be?

Life's Greatest Mysteries

As elements of the human experience, sexual reproduction and death are as inevitable as taxes, and far more enigmatic. From the perspective of science, why should individuals engage in sexual as opposed to asexual reproduction when they not only must expend time and energy in finding a mate, but also dilute by 50 percent their genetic contribution to each offspring? And why should individuals senesce and die when natural selection seemingly would favor any genetic predisposition for greater longevity and continued reproduction? From a theological perspective, the paradoxes are no less profound. Why did a god create humans of separate genders, with all of the moral mine fields that entails? Why would a god sentence us to old age and death with a certainty that trumps every other consideration, including apparent goodness of the individual? Evolutionary science and religion both have provisional answers to such questions, and some of these are closer in spirit than might be supposed.

An essential truth in several religious traditions is that life and fertility beget individual demise, such that life is, in effect, defined by reproduction and death. The Nup people of Nigeria have a creation story that explains the inevitability of this relationship. In the beginning, God created tortoises, humans, and stones, each in the two genders of male and female. At first these individuals could not reproduce, but rather merely became young again periodically. Then the tortoises and humans each decided they wanted children. God said that with reproduction would come death. Nonetheless, they insisted, and God finally acquiesced. Sure enough, after producing children, the tortoises and humans senesced and died. The stones saw what had happened and decided not to make the same request of God. Thus, unlike the tortoises and humans, stones neither make children nor die.

Evolutionary geneticists likewise view organismal reproduction and death as intertwined phenomena, although this perception stems from the logic and operation of scientific mechanisms. By the end of this chapter, these connections between sex and death may seem less mysterious, though no less profound.

Sexual Reproduction

In the *Symposium*,[33] Plato's Aristophanes suggested that the world first was populated by perfect beings who were female on one half of the body, and male on the other. An angry god (Zeus) then sundered the two sides, who ever since have sought to restore their wholeness in love. Aristotle carried the concept further, explaining that the female gender was inherently cold and the male hot. Thus, when warm winds blew most conceptions were of males, whereas cold winds produced female embryos. In the Middle Ages, a popular explanation for unintended pregnancies was that little devils sat on women's chests at night and wickedly impregnated them. Such were the levels of explanation still in vogue for sexual reproduction as recently as two hundred years ago. In the mid-1800s, discoveries of the cellular basis of life, Mendelian inheritance, and Darwinian mechanisms of evolution provided a more modern understanding of sexual reproduction.

In most evolutionary definitions, sex is synonymous with genetic recombination. Usual components of the process are physical recombination (breaking and reuniting different DNA molecules) and outcrossing (mixing DNAs from separate individuals). There are many explanations of why humans have sex. Sexual union is necessary for procreation: One haploid gamete from a male and one from a female must come together to produce offspring.[34] Sex helps to cement the bonds between parents in nuclear families. And, because sex is necessary to initiate reproduction, we've evolved to enjoy it. However, these kinds of explanations are hardly sufficient to explain why reproduction in most higher organisms so often entails only sexual mechanisms.

Certainly for most species, short-term disadvantages attend sexual reproduction. There is the matter of the extraordinary time and energy spent in finding a partner, and of the expenditures and dangers in the often elaborate courtship rituals and in mating itself.[35] In the genome, favorable combinations of selection-tested alleles in the parents might be disrupted by genetic segregation and recombination during progeny production. Sexually reproducing individuals also diminish their genetic contribution to each offspring (compared to asexual reproduction) by 50 percent. These last two factors contribute to the cost of meiosis: the immediate genetic price of doing sexual business.

There are several advantages to sexual reproduction: the possibility of circumventing Muller's ratchet (the ever-increasing genetic load expected in strictly asexual lineages); an opportunity to incorporate into a lineage beneficial mutations that had arisen in separate individuals; and the generation of extensive genotypic diversity among progeny, and with it an added potential for the adjustment of populations to environmental challenges. The major difficulty in proving the superiority of sexual reproduction is in deciding (as natural selection must) how to weigh short-term versus long-term reproductive benefits. The unquestioned fitness costs of sexuality are immediate and high, whereas most of the benefits appear to be postponed and diffuse, though nonetheless crucial to a population's continued survival. This is the central evolutionary paradox of sex. How has an evolutionary process guided by a short-term force, the reproductive fitness of the individual, achieved an outcome (sexual reproduction) whose most obvious proceeds appear to be deferred?

There is little doubt that historical legacy has locked humans and many other higher animals into mechanistic modes of sexual reproduction that make easy reversions to asexuality difficult, regardless of how useful asexuality might be in the short term. Furthermore, species that *have* been able to capitalize upon short-term fitness profits by reverting to strict asexuality typically do not last long, and for this reason also will be underrepresented in extant samples. Thus, in discussions of selective regimes that promote sexual reproduction, most researchers carefully distinguish the genesis of genetic recombination from recombination's evolutionary maintenance.

How did sexual reproduction begin? One hypothesis is that recombinational sex started as an outcome of selection on parasitic DNA sequences that "imposed" biparental sexual reproduction on host genomes to favor their own spread.[36] Recall that selfish DNA elements can increase in population frequency by replicating and dispersing themselves across germ-cell chromosomes, but only when their hosts reproduce sexually. Perhaps recombinational sex arose early in life through selection on DNA-level parasites or mobile elements that were invading host genomes. Another hypothesis is that sexual reproduction evolved through selection on host genomes in favor of recombinational mechanisms for the correction of genetic errors.[37] These and related scenarios, which are not mutually exclusive, often view sexual reproduction in higher animals as a ghost of selection past, from a time early in evolution when recombinational processes arose and the now-familiar mechanisms of sexual heredity became irrevocably ensconced in our genes.

Yet, perhaps short-term advantages to sexual reproduction also exist in higher animals. Many researchers have theorized, with varying degrees of persuasiveness, that the immediate profits from sex indeed do cover the fitness costs to individuals.[38] The arguments typically envision intense "diversifying selection" via environmental heterogeneity that presumably calls for added flexibility in genetic response. Such flexibility is characteristic of sexual more than asexual reproducers. No audit of the fitness books, they contend, would be complete without adequate accounting of these current fitness revenues.

Scientists have envisioned two classes of scenarios. Models of "tangled bank" (the term is taken from Darwin's closing paragraph in *On the Origin of Species*) emphasize how spatial variation in the environment may favor the genetic variety displayed among sexually produced progeny. Given a patchwork quilt of selection pressures, sexual offspring may be more likely to find appropriate niches and thereby outperform asexual offspring, on average, in nearly every generation. Scenarios that take a similar view but emphasize temporal variation in the environment are known as "Red Queen" models. This term is from Lewis Carroll's *Through the Looking Glass,* in which the Red Queen has to run as fast as she can just to stay in place. The analogy is that biotic environments forever are deteriorating from an organism's perspective because of the continual evolution of competitors, predators, and parasites, so that populations must evolve rapidly merely to avoid extinction. For humans, spatial and temporal heterogeneity in the environment is provided by an assortment of parasites and pathogenic microbes. These biological disease agents no doubt play a major role in the maintenance (as well as in the origins) of sexual reproduction by rewarding the recombinational genetic variety that makes their hosts moving targets for exploitation.[39]

Another hypothesis for how sexual reproduction might continue to pay fitness dividends centers on the topic of DNA damage and repair. In an elegant theory advanced by Carol and Harris

Bernstein, damages to genetic material are a universal problem for living things.[40] These DNA injuries are known to be of many types, and they arise from environmental attacks both inside and outside the organism. For example, ultraviolet radiation and many chemicals damage DNA, as do oxygen radical molecules generated inside cells (notably, within mitochondria) as a molecular byproduct of cellular respiration. Remarkably, tens of thousands of DNA lacerations arise in each cell every day! These molecular wounds, if unrepaired, interfere with gene transcription and DNA replication, and can cause progressive impairment of cellular function and eventual cell death.

Fortunately, evolution has produced entire subcellular medical industries devoted to the identification, treatment, and repair of these molecular lesions. A host of intracellular clinicians (proofreading enzymes) and surgeons (repair enzymes) diligently diagnose and mend a variety of genetic traumas, most of which arise during DNA replication. Many of the molecules involved in DNA repair also participate regularly in other cellular activities such as gene regulation, DNA replication, and gene shuffling. Not surprisingly, when the genes encoding such proteins themselves go mutationally bad, the consequences to the cell and to the organism can be disastrous. For example, mutations in DNA repair genes are known to be responsible for several common hereditary cancers, including xeroderma pigmentosum and hereditary nonpolyposis colon cancer.[41]

The intracellular procedures employed by the DNA repair enzymes typically involve rebuilding the damaged DNA, using the intact information from a redundant copy. One source of redundancy is the complementary strand in double-helical DNA, which serves as a template for surgical repair when damage is confined to a single DNA strand. Another source of redundancy, notably available to diploid organisms, is a backup copy of duplex DNA. Such intact duplex DNA is a necessary template for the rehabilitation of double-stranded DNA damage.

The subcellular procedures for double-stranded surgery, known as "recombinational repair," provide the hypothesized causal link between DNA repair and sexual reproduction. The Bernsteins suggest that all mechanisms for molecular recombination, including meiosis in higher organisms, are evolutionary adaptations that originated *and* are maintained by natural selection explicitly for the functions they serve in the repair of DNA damages.[42] A parent contributes only one gamete to each offspring, and indeed, no other class of cells in a multicellular organism influences fitness quite so directly. Thus, the Bernsteins argue, gametes must be as free as possible from DNA defects, and meiosis accomplishes the task. The admixture of chromosomes from separate individuals that takes place in each generation of sexual reproduction ensures a continuing source of undefiled DNA template against which damages to the homologous duplex are repaired during meiotic recombination. Furthermore, the Bernsteins contend, in diploid multicellular organisms with recombination, outcrossing is favored because it promotes the masking of deleterious mutations. Thus, "DNA damage selects for recombination, and mutation in the presence of recombination selects for outcrossing."

The Bernsteins' hypotheses are intriguing not only because they offer a plausible mechanism of how ongoing natural selection might favor the maintenance of recombinational processes, but also because DNA repair provides an explicit mechanistic connection linking the topic of sex to that of aging.

Aging and Death

Senescence and aging, used interchangeably here, are defined as a persistent decline in the survival probability or reproductive output of an individual because of internal physiological deterioration. In other words, we and other organisms become inherently more fragile as we age. It is important to distinguish the progressive and inherent dilapidation of aging from nonaccelerating sources of injury or death such as lightning strikes, car crashes, or falls from cliffs. Consider an analogy to a huge population of glass test tubes in a science laboratory.[43] Test tubes occasionally are dropped and broken. Assume that all breakage is accidental and has nothing to do with progressive deterioration, such as a thinning of glass through handling. The test tubes will tend to decline in number at a stochastically predictable rate (like radioactive decay) and the population eventually will go extinct. However, in the absence of aging, the probability of breakage per unit time for an old codger test tube is no greater than that for a brand new one, and some test tubes by chance may survive for extremely long periods of time.

Living organisms are unlike our idealized population of test tubes. Beyond a certain age typically associated with the onset of reproductive capacity, the probabilities of death tend to increase. Consider girls and women in the United States. The lowest per capita death rates are for ten- and eleven-year-olds, but after that, mortality rates double about every eight years.[44] For instance, the average risk of death for a woman of sixty-eight years is twice that of an average sixty-year-old, and a hundred and twenty-eight times that of an average twelve-year-old! Such demographics have not escaped the notice of life insurance companies, which typically double the rates for approximately every eight years of advancing age.

The marked acceleration of death probabilities with age also explains why there are no ancient humans alive today.[45] Hypothetical two-hundred-year-olds would have a death rate about ten-million-fold higher than that of our twelve-year-olds. By contrast, if a fountain of youth existed such that the death rates of twelve-year-olds remained in effect forever, we would live on average about 1,200 years, and about one person in a thousand today would have been born near the end of the last Ice Age, about ten thousand years ago!

The problem of aging becomes more agonizing when we appreciate that living organisms, unlike test tubes, possess evolved capacities for self-repair. Indeed, such capacity is almost a defining criterion for life. Repair occurs at many levels, ranging from the physiological mending of cuts and broken bones, to immune-mediated recovery from infectious diseases, to the recombinational repair of molecular DNA damages. Given the evident capacity of living creatures to heal themselves, there would seem to be no exigency that individuals age and die. Yet they do. How can natural selection have permitted this state of affairs?

Let us return to the test tube analogy and modify the scenario in a lifelike direction. Suppose a laboratory manager replaces each broken test tube by the purchase of a new young one, and marks the date of replacement on the new tube. When the population of unbroken test tubes is examined twenty years later, many newer test tubes will populate the laboratory simply because the older ones had a greater cumulative opportunity to break. Imagine now that a manufacturing defect makes test tubes of a particular age extremely fragile. The lab manager would hardly notice if the fragility were confined to old test-tubes, because few would reach the affected age anyway. However, a manufacturing defect that made young test tubes fragile would be apparent quickly, and the lab manager probably would contact the supplier for free replacements. The evolutionary point is that natural selection is more likely to eliminate harmful mutations in young individuals than in old ones.

Similarly, senescence and death in real organisms are features that evolved as logical consequences of the declining force of natural selection through successive age classes in a population. The strength of selection on genes in eighty-year-olds inevitably is less than the strength of selection on the same genes in teenagers. Natural selection is more indifferent to the problems of somatic deterioration in old age because these problems are trivial on gene representation in successive generations compared to any difficulties that appear earlier in life. Thus, aging and death exist not because they violate some rule of evolution, but rather because natural selection simply fails to pay sufficient attention to the matter.

Much like the vision of the Nup peoples of Nigeria described earlier, evolutionary geneticists view senescence and death as virtually inevitable repercussions of organismal reproduction. To look at it another way, consider a large, *nonsenescent* population with individuals of various ages, as depicted in the box at the top of Figure 5.3. Because the population initially is not aging, the probabilities (p) of survival from one age class to the next by definition are equal (i.e., $p_1 = p_2 = p_3 \ldots$). If accidental deaths occur and if the population is to survive, individuals must reproduce, thereby generating new members of the age zero cohort (see Figure 5.3, bottom). However, with procreation come age-specific selection pressures favoring reproduction earlier rather than later in life. These selective forces arise because individuals with proclivities for late as opposed to early reproduction have a greater exposure to

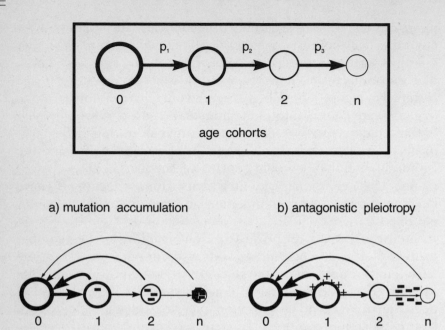

Figure 5.3 Diagrammatic view of natural selection's influence on the evolution of aging. *Above:* mean probabilities (p) of organismal survival through successive age classes within an age-structured population. If the population shows accidental deaths only and no aging, then $p_1 = p_2 = \ldots p_n \neq 0$. *Below:* two alternative hypotheses for how diminishing selective pressures through successive age classes in a population of reproducing individuals can lead to age-specific negative ($-$) and positive ($+$) genetic features characteristic of senescence. In all diagrams, circle sizes are proportional to the numbers of individuals in each age class, and arrows pointing right and left indicate population impacts via relative survival and fecundity, respectively.

accidental deaths and, thus, are likely to leave fewer immediate offspring; and early propagators benefit further when some of their descendants themselves survive to reproduce.

Such reasoning has been formalized mathematically into a population genetics theory that appears to have solved the evolutionary paradox of aging.[46] This theory has received support from laboratory experiments on fruit flies in which researchers forced natural selection to pay greater attention to individuals of old age. In population cages, all eggs produced by flies less than three weeks old were discarded such that only eggs laid late in a fruit fly's life survived to beget the next generation. Under this selective regime, any genes favoring longer survival and age-delayed reproduction were much more advantageous than before, and conversely, any genes promoting early reproduction were selected against. After a mere twelve generations, late-reproducing flies had evolved a 10 percent longer lifespan. The experiments were continued, and flies with life expectancy twice that of the originals recently have been produced! These results are consistent with predictions of the evolutionary theory of aging, and also demonstrate that selectable genetic variation for longevity does exist (at least in fruit flies).

At the penultimate level of explanation for senescence are two conventional evolutionary hypotheses regarding how diminishing selective pressures with age may be translated into "hard-wired" properties of the genome (see Figure 5.3, bottom). The first of these is the "mutation-accumulation" hypothesis,[47] the idea that later age classes become genetic garbage bins where alleles with age-delayed deleterious somatic effects accumulate in evolution because of weak selection pressure there against their loss. As a further aspect of this hypothesis, modifier genes are favored by natural selection that serve to delay the consequences of deleterious alleles to the individual, with the net effect that negative genetic expressions are shoved into older age.

Huntington disease, described in Chapter 3, provides a good example. This genetic disorder of middle and late life is far more common than can be accounted for by a balance between recurrent mutation and negative selection alone. Instead, its prevalence reflects the fact that the gene's horrible effects typically are delayed beyond reproductive age and, hence, are practically invisible to natural selection. Although devastating to the individual, the actions of the Huntington gene are nearly neutral from the standpoint of genetic fitness. By contrast, progeria (a genetic disorder that gives children the symptoms and appearance of old age) affects reproductive fitness profoundly and, thus, *is* strongly selected against. As expected, the incidence of progeria is vastly lower (one in several million births) than that of Huntington disease.

The second evolutionary hypothesis of aging is "antagonistic pleiotropy,"[48] the idea that alleles for aging are favored by natural selection because their beneficial effects at early stages of life outweigh antagonistic deleterious effects later on. For example, genes predisposing for the calcification of bones in adolescents might improve an individual's mean genetic fitness by strengthening limbs, and hence would increase in population frequency despite promoting atherosclerosis (calcification of artery walls) in older age. Any gene or combination of genes that promotes this state of affairs will tend to spread through a population simply because younger individuals make a disproportionate contribution to the ancestry of future generations. As stated by Peter Medawar, "A relatively small advantage conferred early in the life of an individual may outweigh a catastrophic disadvantage withheld until later."[49] Or, as George Williams states, "Natural selection may be said to be biased in favor of youth over old age whenever a conflict of interests arises."[50] Overall, both antagonistic pleiotropy and mutation accumulation probably contribute to the aging phenomenon; the hypotheses are not mutually exclusive.

At more immediate levels of physiological mechanism, the conventional kinds of medical explanations for aging finally come into play. If we are not killed first by a plane crash or food poisoning, we nonetheless senesce and eventually die from cancer, heart disease, stroke, Huntington disease, or any of a host of other endogenous disorders. The genetic underpinnings of these age-related pathologies have evolved to a ubiquitous status because natural selection simply doesn't care whether older individuals survive and reproduce.

One ramification is that death is far more inevitable, in fact, genetically predisposed, than the medical community might lead us to believe. Although mean life expectancy has nearly doubled in the United States in this century, virtually all of the improvement can be attributed to better hygiene, antibiotics that combat infectious diseases, better food and water supplies, and other public health measures that make our environments less hostile. There is little evidence that our genetic pace of aging has been altered even one iota. Rather, we seem backed against a wall of longevity that appears nearly insurmountable. A sobering realization is that if the two current leading causes of death in the United States—cancer and heart disease—were *totally* eliminated, only about six years would be added to mean life expectancy in this country. The marginal dividends to be expected from beating back other genetic disorders of old age will be far lower. If real breakthroughs in gene-based longevity are to be achieved, they will have to come from insights and treatments into the underlying nature of the aging process itself, rather than from combatting individual disorders.

One such general mechanistic possibility for aging returns us to the topic of DNA repair. Recall that in the Bernsteins' view, the inevitable DNA damages that have selected for recombinational repair also contribute to the gradual deterioration of the body's cells.[51] Although refined mechanisms for DNA repair exist in somatic cells, damage control is incomplete and cellular functions are compromised as the molecular insults accumulate.[52] If cells somehow could be insulated against mutational damage, or improved with respect to DNA damage recognition and repair, might the proverbial fountain of youth be tapped? We don't yet know. For the moment, let us look backward at nature's pathways to immortality.

Routes to Immortality

Although no life is truly everlasting, all gene lineages in organisms alive today have survived the replicative transmission process, generation after generation, across the thousands of millions of years that life has been in existence. For all practical purposes, these genes can be viewed as immortal, and it is instructive to consider the evolutionary routes to this astounding achievement. Such considerations also are instructive because they highlight important relationships between immortality on the one hand, and concepts of cellular autonomy and organismal individuality on the other.

Two achievable pathways to genetic immortality appear to have been available to life on Earth.[53] The first is predominantly asexual and is approached most closely by unicellular organisms such as some bacteria. Here, cell proliferation may outpace the rate of accumulation of DNA damages and deleterious mutations with the net effect that Muller's ratchet is circumvented and an indefinite continuance of the population occurs by a cellular replacement process. However, formal models indicate that Muller's ratchet probably sets a strict upper limit on genome size in such asexual organisms. The second pathway to immortality is sexual and is exemplified most clearly by germ-cell lineages in multicellular organisms, such as ourselves, where genomes are vastly larger. Here, mechanisms of recombinational repair operate in conjunction with cellular proliferation and intercellular selection to counter the accumulation of DNA damages and deleterious mutations.

In both routes to immortality, the vast majority of cells (bacteria or gametes) die, but this does not compromise the continuance of cell lineages that happen to have escaped accidental or genetically-based deaths. Thus, the efficacy of both pathways to immortality depends critically on the autonomy of the proliferating cells. To emphasize why this is so, consider the prospect of immortality for a multicellular organism such as yourself. Even if some of your somatic cells and tissues could keep pace with DNA damage through the nonsexual strategy of cellular turnover and replacement, this wouldn't help you survive indefinitely, because the fate of these cell lineages is inextricably tied to the remainder of your soma, which as a whole inevitably senesces and dies (as predicted by the evolutionary theories of aging). In other words, death or severe malfunction of any critical tissue within your body dooms all of your somatic cells, regardless of how healthy they otherwise might be. However, *autonomous* gametic cells and the genomes they contain *can* (albeit with extremely low likelihoods per cell) escape this predicament to live on through your children and grandchildren.

Elements of both strategies—recombinational repair and cellular replacement—are employed simultaneously during gametic production in multicellular animals. The recombinational aspects of meiosis help to purge the nuclear genome of DNA damages and deleterious mutations, and an analogue of cellular replacement at the molecular level facilitates the purging of genetic liabilities in nonrecombining mitochondrial DNA. As mentioned previously, thousands of mtDNA molecules (unlike nuclear genes) populate most cells, and intense intracellular selection is expected to characterize the transmission process of mtDNA from one cell generation to the next. Thus, mtDNA molecules that survive and replicate to populate a mature oocyte, or egg cell, have been screened scrupulously by natural selection for replicative capacity and for functional competency in the germ-cell lineages they inhabit. In other words, molecular replacement operations that apply to asexual populations of mtDNA molecules within cells are more similar to cellular replacement processes in asexual bacterial populations than they are to recombinational aspects of molecular screening that apply to nuclear genes.

The immediate effect of these collaborative processes of recombinational cleansing for nuclear DNA and molecular turnover and replacement for mtDNA is to increase the probability that at least some gametes meiotically produced are "cleaned" of genetic defects that had been inherited or had accumulated during the lifetime of the parent. If these damage-control processes fail during gametogenesis, the metabolic functions of germ cells are compromised and the gametes die. Subsequent rounds of selective screening at the zygotic stage and during fetal development further cull genetic flaws from the genome as surviving gametes that unite to form a new individual are called upon to interact functionally during embryogenesis. These early hours and weeks in the human womb are periods of heavy mortality. "The miracle of life" aptly summarizes the process by which genes have overcome the many selective hurdles in their continuing evolutionary quest to perpetuate themselves through each new generation of multicellular individuals.

Few topics are more fundamental to the human experience than sexual reproduction, aging, and death. They are central to religious involvement, moral perceptions, and societal attitudes.[54] Few subjects so occupy our thoughts or evoke our deepest aspirations, fears, and wonder. Reproduction, senescence, and death often seem ethereal, intangible, spiritual. Few topics have been more shrouded in mystery, cloaked in folklore, and veiled in ignorance.

Yet, in the last 150 years, the biological sciences have moved the topics of sexual reproduction, aging, and death firmly into the realm of mechanistic understanding. Geneticists and cell biologists can describe the amazing specifics of meiosis, fertilization, and development. Physicians can tell us the physiological and metabolic bases of our ailments of age in great detail. Evolutionary biologists have begun to explain the selective reasons for *why* these biological phenomena exist so universally. The currently available genetic models of sexual reproduction and aging do not prove that the evolutionary theories are correct (that must be decided empirically), but they do demonstrate that the operation of natural selection on the genes is consistent with the evolutionary appearance of these otherwise enigmatic and counterintuitive biological phenomena. The sufficiency of mechanistic explanations for sex and aging suggests that the evolutionary-genetic sciences hold a pre-eminent claim for inclusion in any broader intellectual discussions of these most meaningful of human affairs.

SIX

Genetic Sovereignty

I consider ethics to be an exclusively human concern with no
superhuman authority behind it.

Albert Einstein, in *Science, Philosophy, and Religion* (1941)

In this century, the triumph of the scientific method in illuminating molecular aspects of particular human physiologic and metabolic features has been little short of astounding. Less than ninety years have elapsed since Sir Archibald Garrod's prescient suggestion that "bodily form depends on chemical structure," yet today's medical treatises detail the genetic and biochemical etiologies of thousands of human metabolic conditions. Paralleling these sweeping discoveries of cellular and molecular mechanisms have been major conceptual advances concerning the modes by which natural selection, mutation, genetic drift, sexual reproduction, and other evolutionary genetic processes govern the dynamics of genes. The dominion of the genes clearly includes pervasive influence on our bodily conditions. Yet, when it comes to more "ethereal" human qualities such as intelligence, behavioral disposition, psychological profile, mode of social intercourse, personality, emotion, and ethical sensibility, the extent and even the meaning of genetic influence is fiercely debated.

Genes underlying neurological functions and behavioral dispositions evolve by natural selection just as do those for anatomical and physiological traits. Biologists are well aware that not only physical features but similarities in inherited behavior unite members of a species (and often higher taxonomic units). Adaptive coevolution between a species' behavior and its morphology, physiology, and ecology explains why we don't observe vegetarian

139

lions or predatory gazelles. Furthermore, if lions had societal codes of ethics, wouldn't these surely reflect an evolved sense of the rightness of predatory behavior, and wouldn't gazellean ethical standards consider grazing as morally proper and meat-eating as sinful?

Perhaps the evident associations between behavior and ecology in the animal world account for the allure of the belief that human social behaviors and even moral perceptions are ensconced in our own evolution-molded genes.[1] However, to raise such issues on human nature is to sail between the Scylla of full genetic determinism and the Charybdis of pure cultural influence. Where lies the proper channel for safe passage between these flanking dangers? Does this traditional distinction between nature and nurture even provide a proper framework for intelligent discussions on human behavior and ethics?

Toward the nature end of the philosophical continuum is "genetic determinism," parodied in Figure 6.1. Two aspects of genetic determinism can be distinguished. According to some geneticists and social scientists, behaviors that vary among individuals or ethnic groups are those for which genetic differences are best ascribed. According to others, behavioral characteristics nearly universal to the human species provide surer signatures of evolution-molded genetic influence. Toward the nurture end of the debate is "cultural determinism," which interprets both the varieties and the universalities of human behavior as outcomes of societal influence and asserts that most human behavioral tendencies are not in our genes.[2] How can such opposing views be retained in the light of scientific investigations into human behavior? Also, how do scientific views on genetic or cultural determinism of human nature compare to theological stances? Such thorny questions are the topic of this chapter.

Genetic Determinism and the Cyril Burt Affair

Few topics in biological determinism have been more controversial than that of the measurement and interpretation of variation in human cognitive ability. In 1909, a young British scientist named Cyril Burt published the first in a series of highly influential papers promoting the notion that variation in human intelligence was

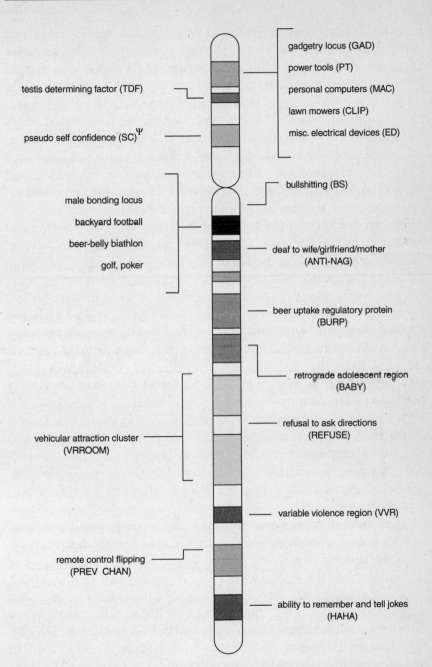

Figure 6.1 Some of the major human genes that have been associated (jokingly!) with the male-determining Y chromosome [modified from *Science* 261: 679 (1993)].

genetically based and heritable. Burt's most convincing evidence came from twin studies in which high correlations were reported in measures of IQ (intelligence quotient) for large numbers of genetically identical twins who had been reared apart since birth. The (questionable) assumption underlying such studies is that a marked resemblance of IQ scores between members of these monozygotic sets of twins must bespeak the influence of the identical sets of genes that they share. Apparent corroboration came from further reports by Burt of correlations between IQ and degrees of genetic relatedness between more distant classes of relatives, such as grandparents and grandchildren, or uncles and nephews. Sir Cyril Burt achieved great fame in his lifetime, as evidenced by his knighthood and by numerous laudatory references in psychology textbooks to his compelling body of scientific evidence for genetic influence on variation in human intelligence.

However, subsequent developments led to vastly different interpretations of Burt's legacy to science. Some scientists had harbored reservations about questionable details of Burt's research findings, but it wasn't until a painstaking biography appeared nearly a decade after his death that his extensive scientific misconduct was documented.[3] Over the last thirty years of his career, Burt apparently fabricated data, conjured up fictitious research associates, invented a student thesis, and in general perpetuated a whopping fraud upon an uncritical if not downright gullible scientific community all too eager to accept the tenets of genetic determinism. The Burt episode now is remembered as a sorry debacle in the history of science, and it casts a lingering pall over all genetic research into human behavior.

Cyril Burt had been born into the British upper class at a time when progressivist arguments espousing biological determinism often were used to support the concept of Anglo-Saxon genetic superiority and the manifest destiny of Britain to rule the world. Eugenics movements, preoccupied with the biology of race and social class, were gaining momentum in England and the United States as supposed means to improve the human species genetically. Into this social climate, Cyril Burt in Britain and Lewis Terman in the United States introduced a modified version of an intelligence test invented by Alfred Binet in 1905. Binet's test had been designed with the intent of identifying children whose performance lagged behind the norm and who might profit from special tutorial efforts. Binet railed against suggestions that results of his test reflected innate differences among children, calling such interpretations "brutal pessimism." However, in the hands of Burt and Terman, IQ test scores soon were interpreted as indices of the genetically fixed intelligence of their bearers. Such interpretations fueled self-serving rhetoric about the genetic superiority of particular individuals and races, with predictable calls for programs to curb prolific breeding by the genetically inferior.[4]

It is profoundly ironic that the cultural environment in which Burt was raised appears to have had such overwhelming influence on his convictions about genetic determinism. In a recent twist to this bizarre melodrama, two independent researchers have examined the records again, and concluded that some of the accusations of fraud against Burt were themselves exaggerated or fallacious, perhaps motivated by critics' professional and ideological antagonism to genetic determinism.[5] Wherever the historical truth may lie, a strong message comes from the Burt episode: societal and ideological persuasion can prejudice the practice and interpretation of "objective" science.[6]

A broader point should not be lost, however. Science certainly is not alone with regard to the dangers of ideological context. If the findings of honest, conscientious scientists are to be discredited because of their potential cultural prejudices, then to an even greater degree must be condemned any culture-driven writings of theologians that discourage dispassionate reason and objectivity. As an epistemological approach, science stands apart from most other human endeavors as a valiant attempt, however flawed in practice, to interpret phenomena objectively and critically.

Limitations of Psychometric Measures

Many of the complex behavioral attributes commonly discussed in human sociobiology, including cognitive ability, are difficult to characterize for reasons beyond observer bias. Much effort has been expended on the development and calibration of psychometric tests, but questions remain. For example, what do IQ tests actually measure (or mismeasure)?[7] The circular response, "intelligence is what an intelligence test gauges," begs the question of how IQ scores are to be validated by independent criteria and culture-free standards. Surely an encompassing description of human mental capacity implies more than can be captured by a standard exam, regardless of how carefully designed.[8] True, empirical scores on IQ tests are fairly good predictors of human performance in tasks related to the tests themselves, such as schoolwork, and in this context IQ rankings might be put to useful service (as Binet intended) in identifying students for special tutoring. However, history records that eugenicists often had more nefarious goals in mind. Furthermore, many other aspects of human intelligence not well measured by test-taking ability undoubtedly exist.

Measurements and interpretations of other complex personality traits are no less daunting. Can meaningful quantitative scores be placed on a person's aggressive tendencies, altruistic persuasions, motivational level, or moral fabric? Psychometric assessments pose greater challenges than those in the identification of evident physical conditions such as Huntington disease or the color of a person's eyes. These psychometric hurdles only compound the research challenges of identifying the hereditary and environmental components of behavioral variation.

Limitations of Twin Studies

Three types of twin protocols are employed in scientific studies of genetic versus cultural influences on human behavioral characteristics. The least critical approach analyzes the statistical correlations in measured traits between monozygotic (identical) twins reared apart (MZA twins). At first glance, resemblance (i.e., positive correlations) between MZA twins suggests genetic influence. However, such observations need not be inconsistent with substantial environmental control for two reasons: twins share a prenatal uterine environment of potentially profound developmental importance, and positive correlations may exist also in the postpartum rearing environments, particularly when the adoptive families of MZA twins have similar social or economic backgrounds.

A scientifically improved (but highly unethical) experiment of this type would be to separate large numbers of monozygotic twins at the two-cell stage, implant the developing embryos into the uteri of unrelated women, and later expose the children to rearing situations statistically randomized across the full spectrum of societal conditions.

A second category of twin studies attempts to ameliorate these difficulties by comparing the correlations of traits for monozygotic versus dizygotic (nonidentical or fraternal) twins, the latter usually matched for same sex to avoid any biases associated with gender. Many such studies report considerably higher IQ correlations in the monozygotic twin sets, as would be predicted under models of genetic influence. However, one plausible caveat is that the environmental influences to which monozygotic twins were exposed may have been more similar on average than those for dizygotic twins, in which case all nature/nurture bets are off.

A third variation on twin studies involves comparing monozygotic twin-sets reared together versus those reared apart, the rationale being that significantly greater trait differences between separated twins must document environmental effects. A surprising result from this approach is that environmental influences on several psychological-behavioral features, including IQ, often appear minimal.[9] One plausible hypothesis is that genetically dependent proclivities of temperament and personality in the twins, regardless of how they are reared (within limits), cultivate cultural settings that subsequently impinge on these twin's developing psychological profiles. Such temperaments might involve a disposition to socialize, pick fights, or be reclusive. Active and adventurous toddlers, for example, probably elicit different responses from adults than do toddlers who are sluggish and timid; and individuals with a strong positive or negative outlook may tend to attract others of similar ilk. To a considerable extent, each genetic individual *makes* the world to which he or she is exposed. This conjecture emphasizes the important point that genes in effect encode important aspects of a person's extrinsic as well as intrinsic environment. If so, gene-environment interactions severely blur any nature/nurture dichotomy.

Heredity versus Moldability

Another sobering lesson is that even the best documented of heritability estimates[10] carry little or no predictive power for human responses to varying cultural circumstances. For example, even if all variation in IQ scores within a population proved attributable to genetic variety (100 percent heritability), a cultural change nonetheless might alter everyone's cognitive performance dramatically, for better or worse. Thus, high heritability does not imply unchangeability. Conversely, even if all IQ variation in a population proved to be the result of cultural influence (0 percent heritability), social environments not yet considered might unmask the expression of genetic differences that were camouflaged in the monitored setting.

A related point is that estimates of heritability within populations say nothing whatsoever about the causation of differences between populations. Imagine that significant fractions of the variation in IQ *within* groups of white-collar and blue-collar workers were attributable to intragroup genetic variety. In principle, any mean difference between the groups nonetheless might stem entirely from differing educational or cultural backgrounds. Failure to appreciate the distinction between the concept of within-group heritability and between-group differences has been a major impediment to rational discourse about the ramifications of ethnic differences in IQ or other personality traits. Worse, such misunderstanding no doubt has promoted eugenicist propaganda and racist sentiment.

Genetics and Human Behavior

Neither the history of ideological posturing in nature/nurture debates, the potential dangers for societal abuse of heritability concepts, nor the many technical and logistic difficulties of studying complex human personality traits appear to have dampened enthusiasm for research into the genetic and environmental bases of variation in human behavior and psychology. A burgeoning scientific and pseudo-scientific literature includes new reports almost weekly of genetic influence on all aspects of human nature, from neuroses and psychoses to religiosities and ethical persuasions. Few serious researchers expect that complex human behaviors or behavioral disorders typically will conform to the so-called "OGOD" model: "one gene, one disorder." More likely, most behavioral and psychological attributes will prove to be influenced by many interacting genes in collaboration with environmental exposures. In dispute are the magnitude and nature of the genetic contributions.

Proponents of the idea of strong genetic influence tend to fall into two opposing camps whose different emphases nonetheless need not be mutually exclusive. One sociobiological camp focuses on traits that vary among individuals or ethnic groups, whereas the other emphasizes behaviors common to nearly everyone.

Human Behavioral Variety

Do you tend to be a happy person? If so, you can count your lucky genes, according to a recent study of identical and fraternal twins in which more than 50 percent of the variation in the "set-point" of human happiness was estimated to have a genetic basis.[11] Do you tend to be anxious, alienated, and nonresilient to stress? If so, you may curse your genetic fate, according to studies of monozygotic and dizygotic twins in which heritabilities greater than 50 percent were reported.[12] Are you musically inclined? If so, sing a song of praise to your genes that early in life may have influenced the developing suite of neuronal connections in your brain in ways that fostered the acquisition of musical skills.

Some of the most intriguing empirical findings on the heritability of human personality have accumulated during the Minnesota Study of Twins, initiated in 1979. Over the years, more than nine hundred sets of identical and fraternal twins, some reared together and some apart, have voluntarily submitted to a battery of psychometric tests, medical exams, and extensive personal interviews designed to disentangle statistically the inborn versus encultured components of psychological variation. An important conclusion has been reached by the authors of the Minnesota twin studies:

genetic factors exert a pronounced and pervasive influence on most human attributes, often accounting for 50 percent or more of the total variation observed for a trait.

To the extent that such studies reliably capture the slippery phenomena of heritability and genetic influence, these conclusions are bolstered by similar findings from independent twin studies conducted elsewhere.[13] The influence of genes has been reported to apply to a host of psychological and personality features (many of which may be interrelated): sense of well-being, mode of reaction to stress, feeling of alienation, avoidance of harm, self-reported altruism, empathy, nurturance, aggressiveness, general and specific cognitive abilities, assertiveness, positive and negative emotionality, and proclivity toward traditionalism.

The authors of the Minnesota twin studies summarized their work and others': "For almost every behavioral trait so far investigated, from reaction time to religiosity, an important fraction of the variation among people turns out to be associated with genetic variation. This fact need no longer be subject to debate; rather, it is time instead to consider its implications." These interpretations run counter to the widely held belief that human psyches and behaviors are almost indefinitely moldable by societal circumstance. They dispute a recent claim that "we emerge from our mother's womb an unformatted diskette . . . Our culture formats us."[14] To the contrary, the twin studies suggest (whether we like it or not) that human behavioral diskettes are preformatted to a significant degree with instructions written in the biological language of genes.

Human Behavioral Commonality

Another conventional class of argumentation for a biological basis of human nature focuses on behaviors that are culturally widespread. Attention to such "omnipresent" dispositions is motivated by the thought that these may be the features most evidently hard-wired in our genes and shaped by eons of natural selection. For example, do culture-free standards exist by which men perceive beauty in women? Yes, according to some evolutionary psychologists, and often in subtle ways that go beyond obvious signals such as good teeth and skin. Recent reports suggest that men of all cultures find women with a waist-to-hip ratio of 0.7 most attractive, a ratio generally associated with youth, and in some clinical studies with women's abilities to conceive.

A new breed of Social Darwinists sees evolutionary causation in such correlations. Doesn't it make sense, they ask, that humans should be programmed genetically to perceive as attractive any features in the opposite gender that signal reproductive potential, and to interpret as unattractive features that signal ill health? Breasts and muscles are appealing, tumors appalling. Other features that men reportedly perceive as beautiful in women include high symmetry between the left and right sides of the face (a purported sign of developmental stability), facial normalness, and gender-pronounced facial features in directions perhaps indicative of high estrogen levels (full lips) and low androgen levels (small lower jaw). To a mate-prospecting male, these and other "feminine" features supposedly telegraph a healthy partner and desirable genes for potential offspring. Likewise, women reportedly show preferences for particular features in men.[15]

In the 1980s the Human Behavior and Evolution Society (HBES) was founded with the aim of fulfilling a Darwinian prophecy of extending to the human behavioral and social sciences the kinds of evolutionary genetic reasoning that have revolutionized other areas of biology. At HBES, an adaptationist perspective has become a paradigm for reexamining human behaviors that formerly were considered from other interpretive vantages (such as Freudianism). At its annual meetings and in HBES's official journal *Evolution and Human Behavior,* formerly *Ethology and Sociobiology,* no topic is too sacred for evolutionary reexamination. Several of the following examples were retrieved from recent issues of that publication. These are intended to exemplify the diversity of topics addressed, and not necessarily the best of the science. Indeed, as discussed later, some of the more egregious of these examples provide parodies of the "science" they espouse. Others, however, cannot be dismissed so easily.

Why do young children often cry at night, enough to rouse their parents? Because, some sociobiologists have claimed, this genetic adaptation tends to delay conception of the next child, a behavior suggested to have evolved because a youngster's survival is compromised by premature arrival of a competitive brother or sister. Why is pornography so ubiquitous? Because, some say, it is a demand-driven media response to a biologically determined male predisposition for polygyny (mating with multiple females), which in turn stems evolutionarily from a lower mean investment in progeny by men than by women. Why do incest taboos exist? Because, supposedly, the low genetic fitness associated with inbreeding has favored the evolution of proximate behavioral mechanisms, often coded in societal rules, to avoid mating with close relatives. Why do pedestrians and mall shoppers regularly aggregate into groups predictable by gender and age? Because, according to one sociobiological account, particular grouping behaviors in our ancestors were adaptive: individuals with high reproductive potential optimized mating contacts and minimized competitive interference by forming small mixed-sex groups, whereas vulnerable children and the elderly gathered independent of gender into larger groups advantageous in foraging and predator avoidance.[16]

Perceived behavioral differences between the sexes over matters of reproduction have provided some of the most fertile ground for sociobiological speculation. The basic idea is that behavioral dispositions underlying mating systems and gender interactions represent evolved responses to selection pressures favoring the selfish genetic interests of individuals, rather than fully harmonic reproductive ventures between males and females. Thus, the ideal reproductive strategy differs between the sexes. Natural selection should tend to push any gene-based mating proclivities of males toward polygyny (exclusive mating with multiple females), and those of females toward polyandry (exclusive sexual consortion with multiple males, particularly if the males could be enlisted to provide care for the female or her offspring). Because these gender-based selective pressures are in opposition, and because it takes two to reproduce, courtship and mating tendencies presumably reflect evolutionary compromises arbitrated by species-specific ecological and biological circumstances. In humans and other mammals with internal pregnancy and extended parental care, two such important circumstances are a greater assurance of maternity for females than of paternity for males, and a far greater investment in the fetus by females. From these relatively straightforward statements has come much sociobiological speculation concerning the "battle of the sexes."

For example, because male reproductive success tends to be limited by access to females whereas female reproductive success tends to be limited by access to resources, it has been suggested that men and women differ genetically in the elicitation of sexual jealousy. Men reportedly become most upset over threats to their sexual exclusivity with a mate, whereas women respond more negatively to loss of a partner's time and attention. In general, inherent differences in the bargaining positions of men and women over matters of reproduction may have prompted selective pressures favoring the evolution of gender-distinctive negotiating skills and moral codes. These may play out today, for example, in the nature of dating games and marital balance-of-power struggles. Even the styles of personal advertisements in newspapers have been interpreted to reflect evolutionary reproductive themes, with men tending to offer resources and to ask for youth and attractiveness, and females offering and requesting the complementary set of qualities.[17]

Year-round sexual receptivity by human females has been interpreted as an evolutionarily forged bonding behavior that serves to elicit continued attentiveness and resource investment by males. Concealed ovulation (loss of estrus) supposedly evolved in humans either because it strengthened intratribal bonds by minimizing disputes over sexual access, or because it broadened female options in copulatory choice.[18] From these have flowed other evolutionary suggestions. Perhaps women menstruate as an evolved defense against uterine infection from oft-present sperm. A women's vagina provides a potential access point for invading bacteria and viruses (sometimes hitching rides on sperm), such that a periodic extrusion of the lining of the uterus to expel infection may have been favored by natural selection in a species with continuous receptivity by females.[19] Inherent reproductive differences between the genders have prompted speculation that other perceived divisions of labor may be influenced by biological mechanisms as well. Some envision, for example, male proclivities for hunting and female proclivities for gathering in early human societies. The guesswork sometimes extends a great deal further, predicting gender biases in behaviors ranging from memory skills[20] to competitive sports to gardening and sewing.

In sociobiological thought, another evolutionary-genetic arena where the selfish interests of associated individuals often come into opposition is the "battle of the generations." William Hamilton and Robert Trivers were among the first to appreciate that natural selection acts in subtly different ways on genes in the parent that influence parental care versus equivalent genes in the offspring. This produces parent-offspring conflicts of interest over the amount and duration of parental attentiveness.[21] The basic idea is that a parent is related equally to each of his or her biological offspring, so natural selection should favor genetically-based parental dispositions to invest similarly in each child. However, each youngster is related more closely to self than to siblings, so natural selection should favor in the young behavior-genetic dispositions to usurp disproportionate shares of parental resources.

In humans and other animals with extended parental care, it has been suggested that a host of behavioral tendencies leading to family strife, including sibling rivalry, arose as evolutionary resolutions of opposing selection pressures. For example, children throw tantrums or display other selfish behaviors that can be interpreted as exaggerated solicitations of parental concern. Such psychological manipulation may work, but only to a point: parents have been selected evolutionarily to monitor the needs of their offspring, but also to detect misrepresentations of those needs. Another suggested nuance is that children may attempt to blackmail parents into providing additional resources by threatening self-destructive behavior.[22]

Parent-offspring genetic conflicts extend even into the womb, the site of the most intimate of human relationships. Although a mother and her fetus have a considerable underlying harmony of fitness interests, their genotypes are not identical, and this opens a window of opportunity for evolutionary discord among the genes. For example, eons of natural selection on genes in the fetus should favor the placental transfer of nutrients to the fetus, although natural selection should also truncate such transfers beyond some maternal optimum. Unlike the parent-offspring battles, which are played out primarily through behavioral acts, mother-fetus conflicts are mediated by chemical signals. Many medical syndromes (for instance, morning sickness), hormonal conditions, and immunological states associated with pregnancy have been interpreted as evolutionary compromises between the competing personal interests of mothers and fetuses.[23]

The list of topics subjected to such sociobiological interpretations is long, but I'll present only a few more examples. Many authors have interpreted tribalism and xenophobia (the fear or hatred of foreigners) as adaptive behavioral tendencies that served humans well over the millennia of evolution in small isolated groups (but which today are of far more dubious value in our interconnected global society).[24] As phrased by Darwin in a rather biblical-sounding passage: "There can be no doubt that a tribe including many members, who, from possessing in a high degree the spirit of patriotism, fidelity, obedience, courage, and sympathy, were always ready to give aid to each other and to sacrifice themselves for the common good, would be victorious over most other tribes; and this would be natural selection."[25]

A revolutionary concept referred to as kin selection, elaborated mathematically in the 1960s, has had considerable influence in the field of human sociobiology.[26] Classic genetic fitness normally is defined as the reproductive success of an individual possessing a specified genotype in comparison to others in the population. Kin selection invokes a broader concept of inclusive fitness that incorporates an individual's personal fitness *and* the fitness of his or her relatives. An individual's genes can profit under the mathematics of inclusive fitness by being transmitted through biological relatives as well as through the individual itself. Thus, a gene encoding a behavior that tends to diminish an individual's personal reproduction nonetheless can be favored by kin selection provided that more than compensatory numbers of the gene's copies are transmitted through the individual's genetic relatives. A net evolutionary profit will be displayed, for example, by any "altruistic" gene that influences an individual to forgo production of a child in order to help two siblings (or four half-sibs, or eight first cousins) each to produce one or more children.

Kin selection theory is a powerful concept that may help to

account for some of the otherwise most puzzling of human behaviors: those involving altruistic "self-sacrifice." Why, for example, is homosexuality prevalent when at face value it would seem counterproductive as a Darwinian fitness strategy? According to some sociobiologists, homophile genes proliferated in hunter-gatherer societies via kin selection when close relatives of homosexuals produced greater numbers of children because of assistance provided by helper relatives.[27]

Altruism is one of four major categories of innate human dispositions (the others being aggression, sex, and religion) discussed in E. O. Wilson's book *On Human Nature*.[28] Few biologists would doubt that many of humankind's strongest behavioral inclinations, including self-preservation and other egocentric tendencies, are the evolutionary outcome of natural selection at the level of individual fitness. Nor would many doubt that dispositions for care of offspring, allegiance to family or tribe, or other forms of helping behavior directed toward relatives are behavioral proclivities that evolved under natural selection and kin selection. But what about heroic acts on behalf of strangers, or the apparent readiness of soldiers and martyrs to lay down their lives for country and religion? Sociobiological theory claims to explain even these most genuinely altruistic of behaviors.

Generosity without hope of personal reciprocation is the hallmark of "hard-core" altruism. According to sociobiological theory, such altruism evolved through kin selection in family or tribal units. Genes for hard-core altruism were favored because of fitness benefits conferred on close relatives through the altruists' actions. In contrast, "soft-core" altruism is ultimately selfish, with the "altruist" expecting reciprocation for his or her efforts. Soft-core altruism arose by selection at the level of individuals, and likely has contributed to the evolution of human proclivities for pretense, deceit about intentions, and other hypocritical maneuvers designed to hide the actor's ulterior selfish motives.

Both of these categories of evolved altruism probably underlie many human emotions and behaviors that find exaggerated and elaborated expression in the modern world. Allegiances may be to football teams, nations, or global religions as well as to family units. The anticipated rewards may be the Super Bowl, medals of honor signaling societal respect, which because of our social nature we cherish highly, or a promised paradise in an afterlife.

By briefly dipping into the stewpot of sociobiological literature I do not mean to suggest that all of the scenarios and hypotheses mentioned are well documented or equally plausible. However, they can serve to illustrate some of the limitations as well as strengths of sociobiological interpretations of human nature.

Reservations about Sociobiological Explanations

One potential danger is the sheer allure of evolutionary explanations for human behaviors. One more example will suffice. Women and other female mammals have evolved strong, genetically-based behavioral dispositions for nurturing their infants, in whom they have both an assurance of maternity and a huge prepartum investment of resources. It is only a small step in logic to assume that male mammals are genetically less inclined to invest heavily in child care (due to evolutionary selection pressures stemming from the uncertainty of paternity and a limited investment in the fetus). A "logical" next stride is to suppose that men are predisposed genetically to polygyny, from which might follow a logical hop that men tend to enjoy the prospects of multiple matings through pornography. A leap of inference could lead to the supposition that if frustrated by social circumstance, men are predisposed toward rape,[29] which in turn might catapult the reasoner to a sociobiological explanation for the necessity of formal cultural sanctions (laws) against sexual assault. Such extended journeys of supposition may begin securely enough, but soon become tangled in a maze of conflicting interpretations. For example, many sociologists view rape primarily as an act of aggression against women that has relatively little to do with sexual frustration. Of course, another train of sociobiological reasoning might link rape to aggression.

A related concern is that it is all too easy to formulate evolutionary scenarios to account for virtually any human behavior. Consider some of the more egregious examples given earlier. Because young children often cry at night, an explanation was advanced that this behavior evolved as a means of delaying subsequent parental conceptions. Suppose instead that human children did not wail at night. This could be interpreted readily as a selection-molded strategy for concealment from nocturnal mega-predators on the Pleistocene savannahs where humans evolved. Or, consider cultural sanctions against incestuous matings, which have been interpreted as logical outcomes of the genetic consequences of inbreeding. Suppose instead that such societal sanctions never were codified. Their absence could be explained easily: incest taboos are unnecessary because humans already possess an instinctual abhorrence of mating with close relatives.[30] At current stages of development and testing, some sociobiological scenarios are little more than just-so stories tailored to account for a given human behavior or temperament. Ironically, sociobiological theory in this sense is weakened as a general scientific hypothesis (one amenable to potential falsification) by its capacity, if anything, to explain everything too facilely.

There are several additional impediments to critical tests of human sociobiological theory.[31] First, the selection pressures that may have molded genetic influence over human behaviors are varied, difficult to specify in detail, and often envisioned to have operated under different physical and cultural environments of the past. More than 99 percent of near-term proto-human and human evolution took place over the last two million years, which constitute the Pleistocene epoch. Selection pressures envisioned to have operated under various Pleistoscenarios are far easier to hypothesize than to evaluate critically.[32]

Second, evolved behavioral dispositions, like evolved physiological features, seldom reflect theoretically optimal solutions to evolutionary challenges. Historical contingencies and idiosyncrasies of mutational and recombinational genetic input often lead to suboptimal evolutionary outcomes that contradict sociobiological hypotheses. Finally, humans are not suitable subjects for manipulative experimentation. This ethical restraint precludes more direct approaches for demonstrating genetic influences on behavior, as for example the selective breeding regimes that have been so successful in genetically imbuing dogs with particular hunting skills, or cows and sheep with docility.

Direct versus Indirect Genetic Influence on Behavior

None of the technical or interpretive limits of sociobiological research implies that the fundamental tenets of such reasoning necessarily are flawed. Current sociobiological scenarios and hereditarian evidence for specific human behaviors are far from secure, but this should not blind us to the possibility that many human behavioral dispositions have evolved under the same natural processes, including Darwinian selection, that have influenced the evolution of our morphological and physiological features.

Precisely how evolutionary forces have played out in connecting genetics to behavior should lie at the real heart of any nature/nurture debate. At one end of the continuum of causal hypotheses is a genetic culturist view: natural selection has favored, and the human genome has evolved to encode, a capacity for flexibility in behavioral response to environmental challenges. Much of this pliability arises from humans' unique mental capacity and the elaboration of culture as a proximate agent of adaptation. At the other end of the continuum of causal hypotheses is an atomist supposition that each category of human behavior is tied directly to particular genes. Both views are evolutionary as opposed to providential, but differ in the interpretation of the genetic program's relationship to behavior.

Epigenetics, Culture, and Phenotypic Plasticity

Epigenetics refers to the entire suite of mechanisms, developmental pathways, and social and other environmental influences by which genomes give rise to organism-level features. For nonhuman species, the usual evolutionary paradigm for interpreting adaptive differences among individuals is to distinguish trait variation due to genetic polymorphisms from that due to phenotypic plasticity.[33] Phenotypic plasticity is the name given to the phenomenon whereby during an organism's development a given genotype can yield different morphologies, physiologies, or behaviors depending upon environmental regime. Humans have evolved an epigenetic agent for behavior so embellished and hypertrophied as to be essentially unique in the biological world: culture. Even if all humans could be identical genetically, they would differ behaviorally to a considerable extent as a consequence of nurtural differences.

Human cultural elaborations, with their deeper evolutionary roots in our genetic endowment of intellectual capacity, have altered fundamental ground rules by which humans adapt to environmental challenges. Through the accumulated knowledge and technologies that culture provides, we have gained the ability to change environments to suit our genes, and thereby have emerged partially from the realm of traditional biological evolution where genes in effect evolve to match environments. These capabilities have enabled *Homo sapiens* to occupy an astounding diversity of habitats and to achieve considerable dominion over other species. A point often has been made that human evolution is now more Lamarckian than Darwinian, in the sense that adaptive capabilities acquired during the lifetime of an individual can be passed culturally to others of the same and subsequent generations, without further direct involvement of the genes. The epigenetic phenomenon of human culture, itself a product of biological evolution, inculcates our species with a degree of behavioral phenotypic plasticity that is unprecedented in the history of life on earth.

The evolution of a genetically-based intellectual capacity for culture serves to broaden a point made earlier in connection with IQ: that human genetic endowments in effect condition physical and social environments to which individuals are exposed and no doubt epigenetically influenced. These interactions again emphasize how the nature/nurture dichotomy can be strained or even fallacious. For example, in the twin studies cited earlier, the fraction (30–50 percent) of variation in IQ (and in numerous personality traits) among individuals that was *not* attributable to genetic influence nonetheless is genetic in the indirect sense that genes predispose learning and culture. Furthermore, to the extent that human cultural practices are a nonrandom draw from a broader pool of cultural practices theoretically imaginable, human behaviors become further tied to the particular genetic dispositions of our species.

From the perspective of comparative biology, human behaviors and cultural practices are strongly circumscribed relative to what is observed elsewhere in the biological world. No human society displays the extreme reproductive division of labor and unflagging selfless devotion to community exhibited by highly social ants and wasps. No antagonism between human females and their offspring (nor soft-core altruism by women) has escalated to the point of matriphagy seen in some spiders whose children routinely eat mom for dinner. No human societies regularly countenance sibling cannibalism, yet the practice is widespread in fishes and many other groups. No human population has dispensed entirely with males as have some fish and reptile species that engage in parthenogenetic reproduction. Few humans are strictly asocial, yet such behavior is the norm for many species. No humans reproduce as functional hermaphrodites, yet the phenomenon is widespread elsewhere in the animal and plant world. Such lists are endless. Imagine the kinds of human cultural practices and ethical values that might be displayed if evolutionary-genetic selection pressures had predisposed our species to these behaviors. Such scenarios may be the stuff of Hollywood fiction, but they are hardly outside the realm of biological possibility.[34]

The broad conclusion is that genes and culture coevolve. Human culture is a product of a highly structured organ, the brain, which in turn is a physical product of genetic evolution. Yet the individual's mind to a considerable extent creates itself through the environments it conditions, continually receiving sensory input and selecting courses of action from among those made available by the individual's social and cultural context. These pieces of cultural information, called culturgens by C. J. Lumsden and E. O. Wilson and memes by Richard Dawkins,[35] include the great variety of creation myths, ethical precepts, marital customs, schoolroom teachings, and the myriad of other societal influences and mental abstractions to which we have been exposed. Nonetheless, these represent only a small and biologically filtered subset of the universe of possibilities, some of which themselves may remain nearly unimaginable to our evolutionarily constrained minds. The phenomenon of cognition (broadly defined) that emerges from the gene-culture interaction *is* the essence of human nature.[36]

The Atomization of Human Behavior

Although genetic influences on many aspects of human cognition are mediated and modulated by culture, one research tradition is to consider the possibility of more direct mechanistic connections of particular behavioral traits to particular genes. Such causal links, if identifiable, would once and for all establish direct genetic contributions for such traits, and also might offer prospects for interventions of the sort described in the next chapter. Initial research has focused on attempts to define and catalogue the unitary behavioral objects to be understood, and then to characterize these at multiple levels ranging from clinical description to neurological function to genetic causation.

One prominent school of thought in evolutionary psychology proposes that the human mind includes a collection of modules selectively designed to solve the most important functional problems for fitness encountered during humankind's evolutionary history. These behavioral modules may be called upon more or less individually when we find ourselves in particular predicaments.[37] Such an arrangement of the mind, it is argued, can better respond to such diverse challenges as facial recognition, child rearing, adaptive social interchange, and language acquisition than would a single general-purpose cognitive arrangement.

A somewhat different possibility is that emotions such as fear, anger, lust, envy, pride, or romantic love are building blocks of human behaviors that correspond to specific situations rather than to specific functions. In this view, genes responsible for particular emotional tendencies have been shaped by natural selection according to circumstances routinely encountered in human evolution. Thus, a given emotion simultaneously may serve multiple functions. Fear, for example, leads to physiological arousal, avoidance and escape reactions, and communication of danger to others. Multiple emotions also unite to serve a single function, such as when fear, envy, and anger combine to prevent loss of a mate. In general, emotions are either pleasurable or painful because of the evolution-forged linkage of these perceptions with fitness-enhancing and fitness-reducing behaviors, respectively. If fear rather than sexual contact had been associated with reproduction, the former would be perceived as pleasure and the latter as pain.

Unfortunately, such broad behavioral modules or emotional descriptions have not yet been connected with detailed molecular or genetic traits. Correlational evidence of genetic involvement for such features has come from the twin studies described above, but in few instances have specifiable genes and their particular metabolic agents been implicated in trait expression at these general levels.

The fields of medical and clinical psychology have provided other compartmentalized descriptions of human behavior for which molecular or cellular causation can be sought. Pronounced mental disorders provide among the clearest examples.

Schizophrenia, the most common psychosis, involves disorders of the thinking process that lead to delusions and hallucinations, paranoia, and withdrawal of the individual from other people and the outside world. Genetic influence is suggested by the fact that relatives of schizophrenics are at several-fold greater risk for the disease (even when raised separately), and that an identical twin of a schizophrenic is twice as likely as a fraternal twin to display the disorder. Evolutionary speculation has centered on why schizophrenia is so common in diverse societies worldwide (about 1 percent of the population), and whether it may have had some selective advantage.[38] However, a clear understanding has yet to emerge. Because of the diverse clinical expressions of schizophrenia and the likelihood that these are due to different kinds of brain damage, schizophrenia now is considered to be a group of mental disorders rather than a single entity. Thus, any genetic contributions to "schizophrenia" as broadly defined probably will not be uniform or simple.

The brain damage responsible for many neurological disorders often is environmental in origin, but the mere fact that such deficits can be localized neurologically suggests that even the most subtle of human perceptual capabilities may be studied at cellular and molecular levels. Prosopagnosia is a neurological condition in which patients can describe human faces with accuracy, but remain unable to recognize at sight even close friends. At least two behaviorally and clinically distinct types of prosopagnosia are known, stemming from lesions in different areas of the brain. Another condition that sometimes entails a specific deficit in brain function is dyslexia, a neurological disorder first noticed as reading and language difficulties in children. A biological marker for developmental dyslexia recently was identified in the form of distinctive brain activity patterns localized to a small area of cerebral cortex.[39] Anosognosia is another example of a condition entailing a specific deficit in brain function, in this case involving damage to the right parietal lobe. Patients are paralyzed on the left side of their bodies but steadfastly refuse to acknowledge the disability.[40]

Disruptions of neurological function occasionally may arise from genetic mutations also. Some forms of dyslexia someday may provide empirical examples because they tend to run in families. The term genocopy has been suggested for any genetic effect that mimics a more widespread phenotypic condition induced by the environment.[41] The word is a reverse counterpart to the more commonly employed term phenocopy, an environmental effect that mimics a genetic condition.

Autism or mindblindness is a devastating neurological disorder clinically diagnosable by the age of three.[42] Autism typically is associated with moderate mental retardation, but some autistic children have normal or superior IQs.[43] What all autistic people share is an obliviousness to other people's thoughts and feelings. By five years of age, most nonautistic children understand that brains are for thinking and dreaming, but autistic children never develop a working concept of their own or other people's minds; they tend to relate to others as opaque, alien beings. Although the behavioral development of autism in infants and children is described reasonably well, neither structural neurological deficits nor genetic alterations have as yet been associated with the disorder. One possibility is that normal genetic operations or cellular factors are disrupted at critical times during an individual's development, but only transiently. Certainly, many developmental genes normally switch on or off at critical times and might malfunction only fleetingly, with nonetheless devastating and continuing consequences.

In the past century, medical researchers have identified more than a thousand neurological syndromes that are inherited through families in relatively straightforward Mendelian fashion.[44] Examples include Batten disease, Wolf–Hirschhorn syndrome, certain forms of Alzheimer disease, Huntington disease, and various other neurological disorders. Researchers now are beginning to home in on the particular genes involved in various conditions of the brain (see accompanying box). It is not surprising that neurological disorders often involve mental illnesses,[45] nor that major disabilities generally

---

Examples of brain disorders for which researchers discovered contributing genes during the year 1997

Epilepsy and mental retardation. A mutant gene on the X chromosome can produce these disorders in women. Male carriers of the mutation do not display the effects but can pass the condition to daughters.

Heroin addiction. Some cases have been associated with particular forms of a gene on chromosome 11 that codes for a dopamine receptor on brain cells. This may be the same gene that influences "novelty-seeking" behavior (see the text).

Parkinson disease (PD). A mutant gene was identified on chromosome 4 that produces misfolded proteins. As the latter accumulate in the brain, neurons die, neurotransmission is compromised, and tremors and muscle rigidity ensue that are the hallmark of some early onset cases of PD.

Boxer brain. A particular form of a gene on chromosome 19 involved with the deposition of a brain protein (beta-amyloid) appears to increase the likelihood that boxers display chronic brain injury from repeated blows to the head.

Obsessive-compulsive disorder (OCD). A gene on chromosome 22 was identified that encodes an enzyme that breaks down excess neurotransmitters in the brain. Forms of this gene that yield less active enzyme appear to be associated with some cases of OCD in men. The disorder, experienced by 1–3 percent of the population, is characterized by a pathological compulsion to repeat certain behaviors, such as washing hands.

Source: After J. Glausiusz in *Discover* 38 (1997): 38.

lend themselves to clearer diagnosis and characterization. More subtle genetic, chemical, and cellular influences on brain function and the mind's "normal" operation are more difficult to pinpoint, yet in the long run should prove to be the most interesting.

Recent reports claim to have identified the first specific gene involved in neurotransmission that leads to variation in a normal human personality trait: novelty seeking.[46] The polymorphic gene *(D4DR)* encodes a receptor protein that at the molecular level binds neurotransmitters related to dopamine. At the behavioral level, the gene appears to influence how quick-tempered, excitable, impulsive, and extravagant a person may be. The gene was identified, and the gene-personality link originally sought, because of the known role of *D4DR* in general neurotransmission. The findings were correlational in the sense that they merely demonstrate an association between alternative forms of the gene and psychometric scores on personality tests. Nevertheless, this research suggests that standard variations in personality traits and emotions may be influenced by specifiable genes for molecular modulators of brain function.

Another example is provided by a recently discovered "chronic anxiety" gene that naturally regulates production of the brain chemical serotonin[47]—the same substance medically targeted by the antidepressant drug Prozac. Reportedly, this gene by itself accounts for about 4 percent of the variation among humans in degree of neuroticism. Subtle genetic influences of this sort might be pervasive. Certainly, a plethora of natural (and artificially synthesized) neurochemicals that influence brain operations and human behaviors are known to medicine and pharmacology.[48]

The human brain is a dynamic network composed of about 100 billion neurons, capable of 100 trillion different connections. About 30,000 to 50,000 human genes are expressed (present as messenger RNAs or proteins) in this organ. A new and rapidly growing enterprise in neurological genetics involves isolating, these transcribed or translated molecules in the brain, and then working backward to identify the genes themselves. To what extent this and similar approaches will expand significantly our knowledge of the mind's operation remains to be seen. Psychologists are beginning to ask whether psychology ever will be the same after the genome is mapped.[49] Sir Francis Crick, a codiscoverer of the double-helical structure of DNA, has gone so far as to urge that the neurosciences should extend into a search for the soul itself![50]

The recent enthusiasm for neurogenetic determinism has hardly gone unchallenged, nor should it. Neurogenetic research during the "decade of the brain" has not yet yielded truly profound results. Neurobiology, psychiatry, and psychology are coming into extensive contact with the evolutionary-genetic sciences for the first time, and must completely reorient themselves. Some recent writings seem naively unaware, for example, of previous nature/nurture battles on related fronts and the hard-won lessons to be learned from them. Any synthetic understanding of the mind's operations and of human behavior will not come from myopically reductionistic approaches that neglect multiple levels of biological, personal, and social causation. However, neither will such understanding come from uncritical holistic approaches alone. As in many areas of biological science, communication among disparate specialties has become crucial to solving the remaining mysteries of the mind.

Providence versus Science

The jurisdiction of genes includes many metabolic and cellular operations of human bodies. These genetic influences play out over time through interaction with the environment to produce the physical states that bestow on us (or deny us) health. We have seen how the genes' dominion extends to the most basic of human experiences: sexual reproduction, aging, and death. In this chapter, we have examined suggestions that the sovereignty of the genes extends even further (albeit to an arguable degree) into the realms of human cognition, psychology, and behavior, again as played out in an environmental context that includes human culture. We also have seen that the natural processes that have shaped our genes operate without ultimate purpose beyond that of molecular self-replication. Yet, from out of these amoral evolutionary processes has emerged a biological species to which moral precepts appear so paramount as to constitute one of humanity's defining characteristics.

Theism

Homo sapiens is a uniquely and profoundly religious species. Is this religiosity itself an adaptive, evolved attribute of our species influenced by the genes? The simple sociobiological answer is "yes." Religiosity served our forbears well in the evolutionary struggle for survival and reproduction, some argue, and as a result natural selection has insured that theistic tendencies are firmly ensconced in our genetically-based mental dispositions. As is true for other human behaviors, both variational and universal aspects of religiosity have been subjects of sociobiological analysis and interpretation.

In empirical studies of twins reared together versus those reared apart, researchers recently employed five different psychometric scales to quantify people's religious interest and propensities for fundamentalism. All scales yielded similar results: "genetic factors account for approximately 50% of the observed variance."[51] Such findings indicate that measurable variation among individuals in religiosity, like happiness, mode of response to stress, cognitive ability, and assertiveness, may be attributable in no small part to variation among individuals in genetic endowment.

Perhaps more compelling is the broader line of sociobiological argumentation that the human predisposition to religious beliefs is a powerful and ineradicable aspect of human nature. In his classic treatise, *On Human Nature,* E. O. Wilson distinguished three forms of selection—ecclesiastic, ecologic, and genetic—that collectively were proposed to have influenced the evolution of human theistic proclivities. Ecclesiastic selection operates via the differential cultural transmission of the rituals, beliefs, and conventions of various religious practices. Ecological selection operates via the screening of these religious practices by environmental demands, as for example when different societies are strengthened or weakened in part according to how their religious and theistic practices dispose them to fare in matters of intertribal warfare, the maintenance of a sustainable social and physical environment, or proclivity to procreate. Genetic selection refers to the evolutionary response of genes to this nexus of ecclesiastic (cultural) and ecological (environmental) influences on human survival and reproduction.

Wilson suggests that "religion is above all the process by which individuals are persuaded to subordinate their immediate self-interest to the interests of the group . . . Incest taboos, taboos in general, xenophobia, the dichotomization of objects into the sacred and profane, . . . hierarchical dominance systems, intense attention toward leaders, charisma, trophyism, and trance-induction are among the elements of religious behavior most likely to be shaped by developmental programs and learning rules. All of these processes act to circumscribe a social group and bind its members together in unquestioning allegiance." All of these proclivities, Wilson suggests, are products of human physiology, ontogeny, and learning, which themselves are features ultimately constrained by the developmental interplay between genes and cultural environments.

Thus, the sociobiological view is that human philosophical tendencies toward religiosity and theism exist because of the Darwinian fitness benefits they conferred upon their practitioners over the course of human evolution. The means by which such benefits may have been achieved are not difficult to imagine. Sacred rituals, rites of passage, and similar religious practices mobilized primitive societies and emotionally empowered their members. Willing subordination to charismatic leaders and group standards added cohesion to societies. Unwavering belief in an omnipotent god and the righteousness of a god's perceived commandments provided supreme motivation and rationale for all manner of "virtuous" actions (almost invariably directed toward the betterment of an individual or his or her group, whether or not at the expense of others).[52] As Wilson notes in *On Human Nature,* "the one form of altruism that religions seldom display is tolerance of other religions. Hostilities intensify when societies clash, because religion is superbly serviceable to the purposes of warfare and economic exploitation. The conqueror's religion becomes a sword, that of the conquered a shield."

Morality

Right versus wrong, good versus evil, moral versus immoral— most of us teach these binary distinctions to our children, attempt to live our lives accordingly, and evaluate others' behaviors by these standards. Candidates for public office tug at our emotions and vie for votes with stances on moral issues ranging from flag-burning to abortion rights. Societies codify norms of morality into laws that sanction acts such as theft or murder, and promote others such as the maintenance of a nuclear family and responsible child care. Religious leaders harp on moral commandments and trumpet the virtuous life. We endlessly measure ourselves and others by moral yardsticks. At the same time, ethical standards are challenged continually and violated by significant numbers of individuals.

Why are humans obsessed with such valuations of behavior and motive, and what are the sources of the evaluation criteria? By now, the general sociobiological answers should be obvious. We are a moderately social species whose evolutionary history has shaped our genetic endowment in ways that have promoted thought processes and behaviors that enhance reproductive fitness of individuals and their kin.[53] According to a recent study, three conditions of biology and natural history led to the evolution of moral perceptions in humans: an individual's dependence on a social group for necessary goods and services, such as food acquisition and predator defense; mechanisms for cooperation and reciprocity within a group; and conflict resolution necessary to preserve the benefits of group living. In human evolution, inter-individual conflict is fundamental, but so too are mechanisms for behavioral restraint and group maintenance.[54]

The behavioral routes to reproductive success involve flexibility, reciprocity, selfish opportunism, and conflicts and confluences of interest. We are attuned to judge the behaviors and intentions of others because the capacities to do so have been within the scope of human biology, and have been favored by natural selection operating with the warp of human intellect and the weft of cultural influences. The richly interwoven fabric that has emerged in evolution includes the human capacity for ethics, and flexible biases for particular moral perceptions and behaviors. Francisco Ayala states, "Moral codes, like any other cultural systems, depend on human biological nature, and must be consistent with it in the sense that they could not counteract it without causing their own demise."[55]

Yet most humans now and in the past have maintained an entirely different view: ethics and morality are absolute rather than biologically relativistic, dictated rather than derived. Some theologians view human nature as inherently evil, with individuals constantly striving to overcome natural tendencies and thereby approach some absolute standard of perfection. Others assume the reverse, viewing human nature as inherently good but continually tempted to sin. Both of these views take ethical codes as givens, with the intellectual challenge being to understand why humans do or do not measure up to the standard. Both of these approaches fail to address the sources of the moral stances themselves. This statement holds even if moral pronouncements derive from a god. The question of source then becomes: Why has the god chosen particular moral commandments as opposed to others?

One of the most difficult questions for traditional theology as well as for science concerns the concept of free will. In one of the biblical stories of Genesis, humans were created in God's image but soon became corrupted by sins in the Garden of Eden. According to some Christian doctrine, all of Adam and Eve's descendants must assume responsibility for their own personal actions; heaven and hell are the eternal rewards for the moral path "freely" chosen in life by each individual. But how could an *omnipotent* and ethical creator hold people responsible for their deeds if he originally created them? Is not God then responsible for all evil?

Such recalcitrant issues simply dissolve under an evolutionary-genetic perspective in which genes and the amoral forces of nature that have shaped them ultimately account for the human condition. Our predilections for developing ethical standards and moral positions have come into being not to satisfy some theistic mandate, not to accommodate some existential blueprint, not to satisfy some universal decree, indeed not for any ultimate purpose whatsoever. Rather, these human proclivities exist because they evolved: they have facilitated the perpetuation of genes across evolutionary time.

However, such scientific understanding alone cannot settle the issue of free will, which is philosophical and definitional. Our evolutionary heritage channels and sets flexible boundaries on human thought and action, such that by definition no exercise of will can truly be free. On the other hand, if there is any transcendental uniqueness to human evolution, if there is any aspect of biology unique to our species beyond cognitive capacity, it surely resides in our unprecedented scope for varied behavioral responses, both as individuals and groups. From this comparative perspective, humans have more free will than any other animal species on earth.

Imagine a species, *Neohomo sapiens,* with twice the behavioral plasticity as our own. Or better yet, imagine visiting spacemen with behavioral flexibility a hundred times greater. Under the variable environmental challenges experienced by these aliens in their intergalactic travels, such behavioral adjustability might well be highly adaptive. Yet according to a strict definition, this species too would lack free will. Whether the behavioral flexibility that humans possess is to be designated as free will is mostly a matter of definition. Whether it is to be interpreted as active adjustability or passive pliability is largely a matter of philosophy. The empirical fact remains that humans can and have created for ourselves many different physical and cultural worlds. With that realization must come some measure of responsibility for our actions in this life, and for our collective future.

SEVEN

New Lords of Our Genes?

Aging and death do seem to be what Nature has planned for us.
But what if we have other plans?
    Bernard Strehler, in J. Lyon and P. Gorner, *Altered Fates*[1]

Within the last 10,000 years, many human societies have shifted in cultural base from small tribal units on natural landscapes to remarkably dense urban assemblages. Automobiles have replaced lions and tigers as a source of predation, business and financial acumen have replaced hunting and gathering prowess as means of resource acquisition, and death far more often than before has been postponed into post-reproductive years. Do such profound changes imply that natural selection has ceased to operate in modern times? Not at all—only the nature of selective pressures has changed. Because human families continue to produce varying numbers of offspring, natural selection (the differential survival and reproduction of individuals and their constituent genes) continues to operate in our species.

Nonetheless, the changing nature of selective agents in industrialized societies is not to be denied. Blurred vision is no longer a serious fitness concern; eyeglasses or corrective surgeries are available. Thanks to modern surgical techniques, anesthetics, and antibiotics, appendicitis attacks or refractory vaginal births are no longer sure death certificates. Measles, polio, or certain other infectious scourges of bygone days no longer hold sway over our lives; vaccines can provide life-long immunity against these plagues.[2] Other changes are less rosy. Environmental toxins and pollutants from modern industry have increased cancer rates dramatically. Many of the earth's life support systems, from clean water supplies to the protective ozone layer in the atmosphere, are

171

under increased pressure from the collective influences and effluents of burgeoning human numbers.

Modern societies have altered the social as well as physical challenges faced by its members, sometimes in subtle ways. For example, perhaps fewer of us can count upon the assistance of extended families in child-rearing because professional demands tend to distance family members in highly mobile Western societies. Whereas large numbers of offspring often were a blessing in labor-intensive agrarian societies, today large families aren't necessarily desirable from an economic or social perspective. In short, physical and social environments have changed, but biological evolution based on fertility and viability differences goes on.

At a pace unprecedented in the history of life, our species has altered environments and thereby modified both consciously and inadvertently the selection pressures to which we (and other species) are exposed. Profound though these culture-mediated developments have been, within the last twenty years the stage has been set for even more unfamiliar modes of evolutionary change. Previously, humans, like all other species, passively awaited genetic change from mutation, genetic drift, natural selection, and other natural evolutionary forces. Now, recent technological breakthroughs in molecular genetics have opened possibilities for conscious and direct human engineering of the genes themselves. With such revolutionary capabilities come profound ethical quandaries and responsibilities.

In this chapter I will introduce some of the developing tools of molecular biotechnology and genetic engineering, provide examples of their actual and contemplated use from the relatively straightforward to the incredible, and consider the philosophical ramifications of these technologies for our relationships to the genes.

Laboratory Methods in the Biotechnologies

Recombinant DNA

In 1968, molecular biologists were intrigued by how microbes, which lack immune systems, protect themselves from viral infection. They discovered a class of biological weapons, produced by bacteria, that shreds invading viruses to pieces. In nature, these "restriction enzymes" snip up viral DNA at specific recognition points determined by short nucleotide sequences.[3] More than 400 different restriction enzymes, each of which cuts DNA at unique sites, soon were isolated by researchers from various bacterial strains. In the laboratory, these molecular scalpels can be harnessed by human genetic engineers as precision instruments to snip DNA from any source at specified nucleotide sites in the molecule (see Figure 7.1). Restriction enzymes provided one of the first and most important sets of molecular utensils in the biotechnologist's toolbox.

Biochemists long have known of another set of enzymes, called

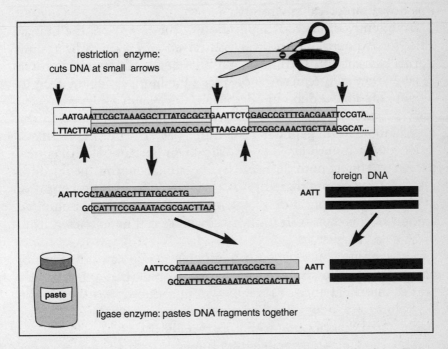

Figure 7.1 The cut and paste operations of recombinant DNA technology. First, a restriction enzyme is used to snip up native DNA from a source such as human tissue. Foreign DNA from another source is also snipped. The snipping operations create "sticky" ends on the restriction fragments. Then, the native and foreign DNA fragments are mixed together in a test tube with ligase enzymes. The sticky ends of DNA clasp one another, and the ligase chemically seals the strands, thereby generating a recombinant DNA molecule.

ligases, that perform the reciprocal task of joining fragments of DNA into longer strands. Just as medical surgeons use sutures and glues to mend the slashes produced by their scalpels, so molecular surgeons now employ DNA ligases in conjunction with restriction enzymes to perform the cut-and-paste micro-operations of "recombinant DNA" technology. Such is the near universality of basic molecular processes across life that restriction enzymes and ligases can be used collaboratively to clip and surgically join genes from even the most distantly related organisms. Indeed, one of the most common procedures in the production of transgenic organisms is to cut a particular gene from the genome of a higher animal, such as a human, and splice it into a bacterium (see Figure 7.2). Quite miraculously, the transgene often continues its normal function in this novel biological environment.

In the first commercial application of this approach, the human gene for insulin hormone was transferred to *E. coli,* a bacterium normally inhabiting the human gut but equally happy when grown in laboratory culture in the appropriate medium. Using the recombinant DNA procedures just discussed, the insulin gene was spliced into the microbe and successfully expressed there, producing the human insulin polypeptide. Then, huge bacterial cultures (fermentation vats) mass produced this medically important pharmaceutical product. This is possible because the human insulin gene, once inside *E. coli,* is multiplied to vast numbers during the course of normal genetic replication by the microbial host. Nearly unlimited amounts of insulin to treat human diabetes can be extracted from these cultures.

The genetic donors in such transgenic manipulations need not be humans. A bovine gene for somatotropin (BST) that greatly boosts milk yield in cows recently was introduced into *E. coli.* The BST hormone, now microbially manufactured, promises to make a substantial contribution to the dairy industry.

Also, the transgenic recipients of human's or other species' genes need not be microbes. One of the first attempts to engineer a transgenic mammal involved transfer of the human gene for growth hormone into mice. Although only a small fraction of the mice accepted the gene and transmitted copies to their offspring, those individuals grew much larger and became known as super-

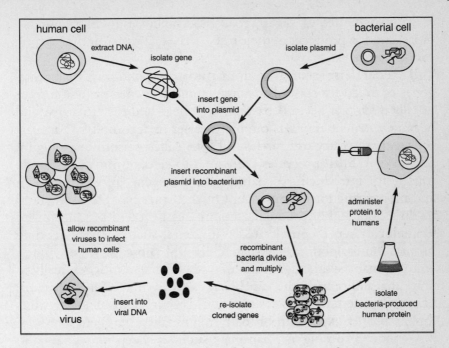

Figure 7.2 Simplified flow chart for two of the major routes to human genetic engineering. Cloning of a particular gene usually is initiated by isolating the gene and inserting it into a bacterial plasmid. The recombinant bacterium then divides and multiplies, producing multitudinous copies of the gene. In some cases, the human transgene within the bacterium may produce a therapeutic protein that can be isolated in large quantities and administered to treat human disabilities. In other cases, the cloned gene may be inserted into a virus that can infect human cells, thereby permitting introduction of the human transgene itself into a patient.

mice. One of the first anticipated commercial successes of this approach is likely to involve the protein α-1 antitrypsin (AAT), which is useful in the treatment of emphysema and of inherited AAT deficiency, a common genetic disorder affecting some 40,000 Americans. Scientists have engineered sheep carrying the human AAT gene, and have purified AAT from their milk in quantities sufficient to suggest that the worldwide human demand for this therapeutic protein might be satisfied by a transgenic flock of as few as a thousand animals. In the near future, expanded applications of such genetic "pharming" procedures might use domestic animals such as cows, sheep, goats, and rabbits as living transgenic bioreactors to produce, in their milk, mass quantities of therapeutic human proteins.[4]

Bacterial fermentation and mammalian pharming of human genes each have technical advantages and weaknesses. The former usually is simpler because bacteria are cheap and easy to grow, and because the gene transfers often are easier to accomplish. Bacterial cells naturally carry tiny circles of DNA called plasmids (see Figure 7.2 on p. 175) that serve as convenient vectors (miniature Trojan horses) for introducing a mammalian transgene into a bacterium and monitoring for its presence. On the other hand, bacteria tend to do a poor job of producing mammalian protein products from lengthier or more complicated transgenes, such as that encoding human hemoglobin. In human cells and those of other higher animals, many synthesized proteins are elaborated biochemically in ways that fall outside the capabilities of the molecular machinery of prokaryotic microbes.[5] Thus, bacteria remain unsuitable as hosts for the proper expression of many mammalian transgenes.

Genetic pharming in mammals such as sheep and goats also has technical limitations, notably the difficulty and cost of establishing a transgenic strain. Several approaches are employed. In one popular technique that requires a keen eye and a steady hand, a fertilized egg first is removed from an animal and, under a microscope, the desired gene is physically microinjected into the egg's nucleus, using a tiny hypodermic needle. The egg then is implanted into the female's uterus. In a small percentage of attempts, the injected transgene integrates into a chromosome of the egg and becomes expressed and heritable, both in the somatic cells of the developing individual and eventually in any of her progeny. Although this approach is used widely, the expense and operational difficulties remain considerable, and a goat which successfully received a transgene can be worth literally tens of thousands of dollars.

A second approach capitalizes upon the infectious properties of viruses.[6] In cut-and-paste procedures similar to those described earlier for bacterial plasmids, a gene in this case is transferred to a virus, which then becomes the genetic carrier that delivers the transgene to mammalian host cells (see Figure 7.2). If all goes well, the gene is expressed properly in its new environment. Viruses have been the molecular delivery systems employed most widely in the early years of human gene therapy. Viruses also serve as a reminder that recombinant DNA technologies (broadly defined) are hardly new on the evolutionary stage. Rather, such recombinant methods were invented by nature hundreds of millions of years ago, and are employed naturally whenever an infectious virus integrates into a host genome.

Additional DNA delivery systems are available to human biotechnologists. Somatic cells such as those in blood or bone marrow can be removed from an individual, genetically engineered by recombinant DNA techniques similar to those already mentioned, and returned directly to the patient. Another technique called electroporation takes advantage of the charged nature of DNA molecules to transfer genetic material across cell membranes under the influence of an electric current. One available gene delivery system, particle bombardment, sounds like something out of Star Wars. In this approach, DNA-coated metallic projectiles, or microbullets, are fired into recipient cells by high pressure air guns that resemble Saturday night specials. Although this method works well for injecting genes into plant cells,[7] the muzzle velocities required have precluded application of this technique to relatively fragile animal cells.

Gene Isolation and Identification

All of the recombinant DNA technologies mentioned thus far begin with a gene to be transferred. How are such genes identified and isolated? One traditional approach, illustrated in Chapter 3 in discussion of the quest for the Huntington disease gene, involves mapping a particular gene to increasingly refined street addresses on a chromosome. Once located precisely, such genes then become candidates for more detailed examination by DNA sequencing or other methods soon to be described. However, the difficult and laborious nature of this conventional mapping approach has prompted the exploration of alternative methods for gene isolation.

One common approach, a bit like an organized fishing expedition, involves the capture of short DNA sequences from functional genes. The key steps in this strategy are to hook random messenger RNA (mRNA) molecules from the pool of molecules in a particular tissue of interest such as the human brain, and then reverse-transcribe them to their complementary DNAs (cDNAs). Recall that mRNA molecules in effect are miniature transponders whose intracellular job is to relay the coded information in particular genes to direct the construction of proteins. A partial reversal of this process, carried out in the laboratory under the auspices of an enzyme known as reverse transcriptase, permits the *in vitro* (outside a living system) recovery of coding portions of genes from their more easily isolated mRNAs. Each stretch of DNA caught in this fashion is known as an expressed sequence tag, or EST. One advantage of this approach is that each EST came from a protein-coding gene, as opposed to the surrounding genomic sea of non-coding "junk" DNA. A disadvantage is that the identity of the gene recovered and its cellular functions are unknown and must be clarified with additional effort. In 1995, a landmark paper reported the isolation of nearly 90,000 unique ESTs, representing a total of 83 million nucleotides, from thirty-seven human tissues at various stages of development. As of October, 1996, more than 16,000 different genes from these and other ESTs had been mapped onto human chromosomes.[8]

The next step is to sequence the isolated pieces of DNA. In 1977, two independent research groups published distinct biochemical procedures for the direct assay of nucleotide sequences in any gene.[9] Over the years, sequencing methods have been modernized. Now, robotic workstations and automated DNA sequencing machines, sometimes housed in huge laboratories with the look and feel of production factories, churn out reams of molecular calligraphy at such a rapid pace that data management by even the fastest computers has become the greater challenge. Within the next few years, an ongoing worldwide effort known as the Human Genome Project is projected to complete the transliteration of the three billion nucleotide pairs that comprise the genetic encyclopedia, or molecular Bible, of our species.[10] This achievement will stand as one of the great milestones in the history of empirical science, nothwithstanding the fact that at least as much or more effort will be needed to understand the full functional significance of these human genomic sequences.

This transliteration is not being accomplished by reading the genome from end to end, as we might read a book, but rather by assembling jumbled bits and pieces of DNA sequence into an ordered whole. These unsorted fragments may be the ESTs just mentioned, or they may have been generated by other recombinant DNA technologies described earlier. For example, restriction enzymes can be used to cleave total human DNA into random but manageable segments, which then individually are cloned to high copy numbers in bacteria, or in single-celled eukaryotes such as yeast. By retrieving and sequencing these DNA fragments, sentences and paragraphs of the human genetic encyclopedia gradually come into view. The genetic phrases must be assembled into the correct order, a task that is not as impossible as it may sound. Many of the DNA sequences, when amplified and sequenced, have extensive stretches of overlap that can be used to align them properly.

In 1983, Kary Mullis, a molecular biologist working for the Cetus Corporation in California, discovered a method of amplifying DNA fragments *in vitro,* thereby circumventing the need for laborious biological cloning through a microbial vector. The "polymerase chain reaction" (PCR) technique permits recovery of assayable DNA from extremely small amounts of starting tissue, in some cases even from a single cell. It also permits scientists literally to bring DNA back from the dead. In Chapter 4, for example, I discussed how PCR recently was employed to recover and assay color-blindness genes from eye tissues that had been saved following John Dalton's death in 1844. Similar molecular genetic methods have been used to recover assayable DNA from the skeletal remains of a nineteenth-century Russian tsar, from Egyptian mummies, from human brain tissues thousands of years old preserved in a Florida swamp, and from bones of a Neanderthal individual who died about 50,000 years ago.[11]

The polymerase chain reaction (see Figure 7.3) generally mimics, with some special twists, the DNA replications that cells perform naturally but far more slowly during normal cell division. In reactions conducted in a test tube, double-stranded DNA molecules first are unzipped (by denaturation at high temperature), and the single strands employed as a template for the enzymatic reconstitution of new replicas of the duplex molecules (at lower tem-

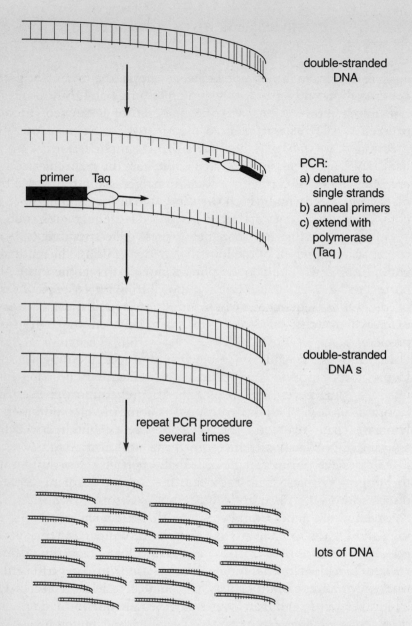

double-stranded
DNA

PCR:
a) denature to
   single strands
b) anneal primers
c) extend with
   polymerase
   (Taq)

primer   Taq

double-stranded
DNA s

repeat PCR procedure
several times

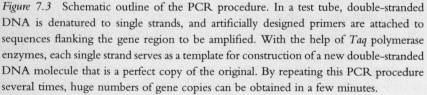

lots of DNA

Figure 7.3 Schematic outline of the PCR procedure. In a test tube, double-stranded DNA is denatured to single strands, and artificially designed primers are attached to sequences flanking the gene region to be amplified. With the help of *Taq* polymerase enzymes, each single strand serves as a template for construction of a new double-stranded DNA molecule that is a perfect copy of the original. By repeating this PCR procedure several times, huge numbers of gene copies can be obtained in a few minutes.

peratures). This process is repeated about thirty times. In each round, the number of facsimiles of the original DNA roughly doubles, and soon there is enough product for DNA sequencing or other purposes.

There are two keys to successful DNA amplification through PCR. First, DNA primers are needed to start the molecular chain reaction, much as a splash of water is necessary to prime an old-fashioned pump. Each DNA primer-pair binds uniquely to short DNA sequences flanking a particular gene region to be amplified. This exactness of binding confers gene-specificity to the amplification process, an important property of PCR. The second key is the availability of an enzyme that both directs the synthesis of a complementary nucleotide strand during DNA replication, and is able to withstand the high temperatures of the DNA denaturation phase of the PCR cycle. One such enzyme, *Taq* polymerase, had been isolated from the bacterium *Thermus aquaticus,* which inhabits the hot springs of Yellowstone National Park. There, at temperatures near 100°C, *Taq* had evolved a remarkable thermal stability that proved to be just what biotechnologists needed for PCR. The discovery of this enzyme also provides a wonderful example of two object lessons for enlightened societal attitudes toward basic research: that findings from pure science often contribute in unanticipated ways to applied technology; and that some of the Earth's rarest and most obscure organisms can furnish fabulously valuable molecular compounds.

Availability of the complete sequence of the human genome will hardly be the end of the story. Decoding the sequences will be the next step toward characterizing particular genes and eventually working out their modes of metabolic action. In the early years, participants in the Human Genome Project debated at length whether research priorities should focus on sequence acquisition *per se,* or whether resources should be devoted preferentially to multifaceted analyses of candidate gene regions suspected of being of special import to human health. In actuality, both tasks have gone forward simultaneously. Thousands of human genes thus far have been characterized at varying levels of genetic and biochemical dissection.

Genetic Diagnostics

The new biotechnologies also provide revolutionary capabilities in genetic diagnosis, where at least two distinct contexts can be distinguished. First, in forensic applications, various molecular assays directed toward polymorphic segments of the genome now are used routinely in utilitarian extensions of Garrod's prediction that each human individual is genetically unique. Such assays find routine applications in courtroom cases where it is of interest to physically link biological material (blood, semen, or hair) to individual victims or suspects whose identity otherwise may be in question. Such molecular methods also are used routinely to establish biological paternity (and sometime maternity) when the natural parent of a child is uncertain.

"DNA fingerprinting" techniques typically rely on polymorphic mini- and microsatellite regions of the human genome. Recall that these loci[12] consist of variable numbers of tandem-repeat (VNTR) families of nucleotide sequences, which reside at many thousands of chromosomal locations. At most such loci in human populations, multiple alleles exist that differ in the numbers of repeat units, and hence in molecular size. These alleles are distinguished by separating DNA fragments through an agarose or acrylamide gel under the influence of an electric current.[13] When many variable loci are assayed simultaneously, the resulting DNA banding patterns on the gels prove to be specific to an individual. The banding patterns look much like the identifying bar codes found on most retail items. The loci also may be assayed one at a time and the data accumulated across multiple genes to provide comparable powers of individual diagnosis.

Whereas DNA fingerprinting methods capitalize upon the variability of genomic sequences with uncertain function to the cell, another class of DNA diagnostic procedures focuses explicitly on genetic markers that themselves either cause human health disorders, or are tightly linked on a chromosome to other genes that do so. The Huntington disease gene again provides an illustration. Identification of the HD gene on chromosome 4 soon led to a diagnostic molecular assay for presence versus absence of the defective allele in any individual. Adoption agencies often require

HD tests for prospective parents with a family history of the disease, so that adoptive children will not have to care for a HD-disabled parent later in life.

This category of diagnostic tools typically involves extraction of DNA from a small sample of a person's blood or other tissue, isolation of the gene of interest, and identification of the gene's alleles based on distinct molecular characteristics of normal and mutant forms as assayed in an electrophoretic gel. If appropriate flanking primers are available, the gene may be isolated by PCR amplification prior to the electrophoretic portion of the diagnostic test. Alternatively, the gene may be revealed by a technique known as Southern blotting that follows the electrophoretic separation of a galaxy of nonpurified DNA pieces. "Southern" in this context has nothing to do with compass heading or geographic region, but rather is the surname of this technique's inventor.[14] When an electrophoretic gel is loaded with chopped-up ("restricted") DNA from many genes, the Southern blotting procedure in effect identifies the alleles of interest via cross-hybridization with a radioactively tagged molecular "probe" designed specifically for the gene of interest. This probe is a copy of the gene previously amplified from another source, either directly by PCR or by biological cloning through a bacterium (as described above under "recombinant DNA").

The field of genetic diagnostics soon may be revolutionized further by DNA chip technology, which promises to extend clinical screening to large numbers of disease-causing mutations simultaneously.[15] The idea is to synthesize short pieces of DNA (oligonucleotide probes) that correspond to each gene under study, and bind them to small squares of wafer-thin glass chips. These microchips then are incubated with the patient's relevant genes, and instances of DNA mismatch (indicating mutations) are monitored by the use of computer software. Microchips have been manufactured with space available for tens of thousands of different oligonucleotide probes. Preliminary trials have indicated the efficacy of microchip technology for rapid genetic screening of conditions such as the presence of drug-resistant viruses in patients with HIV, and particular mutations underlying cystic fibrosis.

This brief narrative tour of a molecular genetics laboratory merely introduces the beguiling diversity of techniques developed in recent years for the isolation and characterization of genes. Before I consider some of the philosophical and ethical questions raised by these technologies, I will present a few additional examples of achieved and contemplated procedures in human genetic engineering.

Genetic Manipulation of the Human Condition

Reproductive Tinkering

Can a sterile man, unable to produce mature sperm, nonetheless father healthy children? Thanks to breakthroughs in human reproductive technology, the surprising answer is "yes." A recent clinical case involved an infertile man suffering from azoospermia, a common condition often of genetic origin (involving, for example, microdeletions in the Y chromosome). Fatherhood was accomplished by a medical procedure, "intra-cytoplasmic sperm injection" (ICSI), which involves the isolation and microinjection of immature, tailless gametic cells (spermatids) from the man's germ line directly into his wife's egg, thereby bypassing many of the complicated tasks normally required of a mature sperm to swim to and penetrate the egg's outer covering. The clinical techniques are reminiscent of those mentioned above in the creation of transgenic animals via hypodermic microinjection of DNA into an egg cell, except that here the injection consists of an entire haploid genome. In the last decade, ICSI procedures[16] have allowed many hundreds of otherwise infertile couples to produce children.

The use of spermatids to achieve conception via microinjection is but one technical nuance within a broader class of *in vitro* fertilization (IVF) methods now used routinely to produce "test-tube babies" for couples or single females desirous of children but unable to conceive by conventional routes.[17] The first test-tube baby, Louise Joy Brown, was born in England in the late 1970s, and since then some 20,000 others have followed worldwide (notwithstanding official condemnation of IVF by the Vatican, and bans on federal funding for IVF research in the United States). Typically, eggs are removed from a woman's reproductive tract, fertilized by sperm in a petri dish, and injected into her womb with the hope that implantation, pregnancy, and successful birth will result.[18] About 10 percent of the time, they do. However, IVF techniques remain expensive, costing thousands of dollars per attempt.

The cell-manipulative technologies being perfected under IVF research (often privately funded) have opened possibilities for other brave new worlds in genetic engineering, including preimplantation genetic screening (to be discussed later), and perhaps the most anathematical prospect of all to many oppositionists: human cloning. It is one thing to clone particular human genes in bacteria or viruses. It is something else entirely to contemplate the artificial creation of genetically identical human individuals.

Nonetheless, such capabilities in essence are available, and already have been employed in the husbandry of domestic animals. Clonal offspring have been produced from managed crosses between bulls and cows, for example. After a zygote has split five times, a miniature embryo is flushed from a prized cow's uterus, and the genetically identical cells (having arisen from mitotic divisions of the zygote) are dissected apart. At this point, each cell is undifferentiated and retains the genetic potential to direct the development of a healthy calf. To realize this potential, the nucleus of each cell is microsurgically transferred to an enucleated egg taken from any ordinary cow, and later introduced to the surrogate womb of a foster mother for embryonic development and eventual birth. By this approach, first successfully conducted in 1986, small herds of genetic carbon-copy cows with desired traits can be created.

In 1997, a mammal was first cloned successfully through the microsurgical transfer of the genome from an *adult* organism.[19] In experiments generally like those described above, a nucleus from a mammary cell of an adult sheep was transplanted to an enucleated egg from another ewe (see Figure 7.4). Several months later, the surrogate mother gave birth to a now-famous lamb (Dolly) who proved to be a genetic clone of her "natural" (or should we say "unnatural") mother. This achievement astounded the scientific world: Most geneticists presumed that the genome from any adult mammalian cell had lost irrevocably, during its own differentiation,

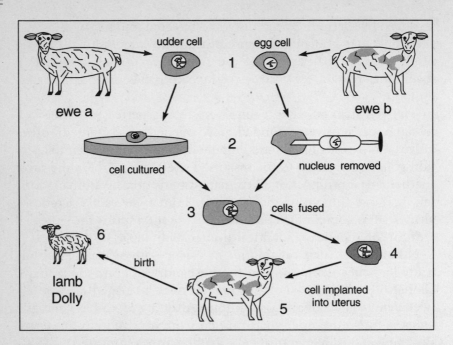

Figure 7.4 Outline of the procedure employed in the first successful attempt to experimentally clone a mammal using the DNA from an adult cell. (1) An udder cell and an unfertilized egg are taken from ewes "a" and "b." (2) The udder cell from ewe "a" is placed in culture under conditions that switch off its genes; the nucleus from the egg cell of ewe "b" is removed by micromanipulation. In (3) and (4), the two cells are placed side by side, and fuse through administration of an electric pulse. (5) The fused cell is implanted into the uterus of a surrogate mother related to ewe "b." (6) A few months later a lamb is born, a genetic clone of ewe "a."

the capacity to direct the full development of an embryo (totipotency). These experiments were shocking in another regard: They raised the possibility that clones might also be created from adult human tissues.

Similar thoughts had been entertained long before Dolly. It was known, for example, that human embryonic cells remain totipotent at least to the four-cell stage, such that genomic transfers to enucleated eggs could in theory be employed to produce clonal basketball squads or dance troups of genetically identical children. Advocates had pointed out discerned benefits. Some parents might desire identical quadruplets. Or, clonal zygotes stored in a hospital freezer could be withdrawn periodically to augment a growing family which then might come to include genetically identical offspring of widely varying age. The stored pre-embryo clones also could provide a genetic insurance policy that guarantees the family a genetically identical child in the event of their child's death. In the uterus of a surrogate mother, production of clonal (as well as nonclonal) offspring could be continued even after the death of the biological parents. It is not too difficult to imagine that surrogate mothers might someday soon give birth to clonal embryos derived from adult as well as embyronic cells.

Genetic Screening

The tools of molecular biotechnology afford many opportunities for the diagnosis of human genetic conditions via DNA typing. In the United States alone, more than 400 clinical laboratories are devoted to the effort. One of the first DNA-diagnostic screens, developed in 1976, was for the common hereditary blood disorder thalassemia. Since then, clinical tests have been developed for hundreds of genetic abnormalities including phenylketonuria, AAT deficiency, Lesch–Nyhan syndrome, sickle-cell anemia, fragile X syndrome, neurofibromatosis, Marfan syndrome, hemophilias, muscular dystrophies, and certain mutations underlying cystic fibrosis. In addition to these molecular tests for inherited genetic disorders, similar DNA–level diagnoses are used to identify the presence of infectious disease agents such as particular viruses and bacteria. Recently, a commercial AIDS kit to detect HIV was made available for use in the home.[20]

The first credited application of mass DNA screening for carriers of a serious genetic condition involved Tay-Sachs disease. Individuals homozygous for the Tay-Sachs allele suffer mental retardation, paralysis, and death usually before the age of five. Heterozygous carriers of the recessive allele are healthy. In the early 1980s, molecular screens applied to more than 300,000 Jewish volunteers worldwide identified numerous carriers of the deadly gene, and such information sometimes was taken into account in subsequent marriage and family plans. The testing program has had its critics, but among the benefits was the virtual elimination of Tay-Sachs disease from Hasidic Jews in Brooklyn.

In principle, genetic testing can be conducted for any genetic disease for which DNA markers have been identified, and at virtually any stage of the human life cycle. Although often described as tests for genetic "birth defects," many of the assays are applied routinely to prenatal individuals. Classic amniocentesis, whereby fetal cells are sampled for genetic diagnosis starting at sixteen weeks of gestation, is used widely to screen for genetic conditions such as Down syndrome in high-risk pregnancies. A procedure known as chorionic villi sampling (CVS) permits the sampling of cells from embryos as young as eight weeks.[21] Early rather than late detection of serious genetic defects *in utero* usually is deemed desirable by all parties concerned because it gives families time to consider their choices.

One of the most remarkable new technologies in genetic testing, pre-implantation genetics, promises to push the temporal diagnostic envelope into the earliest stages of life.[22] With the assistance of PCR, genes can be isolated and then assayed from blastocysts, without harming these small collections of cells, before they attach to a woman's uterine wall. Even more astounding, genes may be screened from individual gametes. For sperm, such characterizations may seem pointless because the technique itself (as currently conducted) destroys the cell. However, for an egg the situation is different. Associated with maturing egg cells, for a brief time, are visible structures known as polar bodies that represent pre-egg cells generated from meiotic cell divisions. Normally, these functionally superfluous cells are jettisoned prior to fertilization. However, they remain present and visible at the time a mature egg is flushed from a woman's ovary during the routine IVF procedures described above. Furthermore, in "polar body biopsy," they can be microsurgically removed without damage to the egg, and their DNA then PCR amplified and genetically screened.

Because they derive from meiosis rather than mitosis, the genomes of polar bodies are not necessarily genetically identical to those of their counterpart egg. In some cases, the genetic composition of the functional egg can be deduced by subtraction. For example, suppose that a woman who is a heterozygous carrier of the allele for cystic fibrosis wishes to have her eggs screened for those that house the normal as opposed to the diseased form of the gene. If two of the assayed polar bodies prove to carry the defective gene, the associated egg cell (and the third polar body) are normal, barring rare *de novo* mutations. An egg that has passed the genetic test then can be fertilized *in vitro,* and returned to the same or another woman's uterus with far better prospects for a healthy baby.

The mere availability of a technique for the diagnosis of a genetic disease does not imply that it should or will be utilized. The diagnostic test for Huntington disease was developed in 1986, yet only small fractions of families at risk for this disorder have availed themselves of the clinical procedure. Many people prefer not to know their genetic fates, especially in the absence of a therapy or cure. Other impediments to widescale genetic screening are economic or logistic. As one leading expert in the field of genetic screening points out, molecular assays applied merely to expectant Caucasian mothers in the United States for just four common genetic disorders would require about eight times more clinical tests than currently can be conducted by all of the nation's genetic centers.[23] Apart from the practical difficulties of mass genetic screening, there are many ethical issues to be considered.

Gene Therapy

To anyone who has tested positive for a serious hereditary disorder, the technological triumphs of molecular biology in genetic diagnosis must seem unimpressive when treatments or cures are unavailable. Unfortunately, the biotechnology horse of diagnostic capability too often has run far ahead of the therapeutic cart. Still, this cart is not empty, and its contents promise to grow dramatically in the foreseeable future.

In 1990, one of the first experiments in human gene therapy was carried out on a four-year-old child with severe combined immunodeficiency (SCID). Ashanti DeSilva was always sick—she spent much of her first four years in quarantine. She had inherited from each parent a defective copy of a critical gene on chromosome 20 that left her body with no capacity to produce the enzyme adenosine deaminase (ADA) required for proper operation of the immune system. A medical team at the National Institutes of Health removed white blood cells from the girl, inserted normal copies of the ADA gene into them using a disabled viral vector, and returned the treated cells to Ashanti's body. Over the ensuing months and years, the girl's condition improved dramatically. Ashanti gradually transformed into a vibrant and healthy youngster.[24] However, this pioneering experiment cannot be claimed as a definitive success story for gene therapy. Only a modest fraction of the girl's white blood cells took up and expressed the ADA gene, and some of Ashanti's improvement probably stemmed from other medical treatments that were administered in conjunction with the genetic therapy itself.

The injection of biological macromolecules into a genetically disabled human is not itself new. Insulin is administered routinely to diabetic patients unable to produce sufficient quantities of this hormone on their own, as are blood-clotting factors for patients with hemophilia. The problem with such therapies is that proteins or other drugs tend to degrade quickly in a person's bloodstream.[25] In Ashanti's case, the revolutionary aspect was the therapeutic injection and expression of a *gene,* and the broader promise that this DNA technology offers for the continued endogenous production of therapeutic compounds by the patients themselves.[26] In essence, successful gene therapy moves a secure production source of therapeutic chemicals from outside of the body to inside.

Similar attempts to replace defective genes by "good" ones currently are in various stages of research development, and more than a hundred clinical protocols involving gene transfer have been approved by regulatory organizations.[27] Among the human disorders currently targeted for clinical trials in genetic therapy are cystic fibrosis, diabetes, Fanconi anemia, Gaucher disease, Hunter syndrome, purine nucleoside phosphorylase deficiency, chronic granulomatous disease, hypercholesterolemia, hemophilia, rheumatoid arthritis, and various cancers. In principle, gene therapies eventually might be directed toward any of the thousands of genetic disorders currently being identified in the Human Genome Project.

For tissue-specific diseases, developing gene delivery systems that target particular cell types to be engineered is a major challenge. For example, retroviruses infect rapidly dividing cells and therefore might prove well suited for the delivery of therapeutic genes specifically to human blood cells or to tumors. Adenoviruses infect cells lining the lung and have been employed in clinical trials to deliver good copies of the cystic fibrosis gene to this respiratory site where they are needed most. Herpes viruses infect cells of the nervous system and are plausible vectors for genes whose products might alleviate neurogenerative disorders such as Parkinson disease. As described earlier, several nonbiological vectors, also in experimental use, avoid some of the potential dangers inherent in the use of mutable viruses. Along with these new techniques come specific technical challenges: the transferred gene must be expressed properly in the cell in which it is inserted, and the transgene must not disrupt other cellular operations.

Gene therapy might also be used to treat or prevent complex, multifactorial diseases. For example, the most common form of occlusive vascular disease, atherosclerosis, typically involves gradual degenerative changes in blood vessels, such as dysfunction or inflammation of the vessels' endothelial linings, or dysregulated interactions with blood cells. Both genes and environmental factors contribute to the development and operation of the vascular system, as well as to variation among individuals in risk factors associated with atherosclerosis: cholesterol processing, diabetes, hypertension, and other physiological conditions that promote oxidative stress.[28] In principle, preventing or treating any of these risk factors through genetic intervention might prevent or alleviate the symptoms of atherosclerosis. The use of *ex situ* (outside the body) microbial fermentation to produce human insulin for the treatment of diabetes already has been mentioned, and *in situ* alterations of the insulin gene within the diabetic's body itself might be feasible. Work has also begun on therapy for several genes involved in cholesterol processing, such as apolipoprotein B and the LDL (low density lipoprotein).[29] For diseases with gradual onset and complex etiology, such as atherosclerosis, it may be difficult to critically document the beneficial results of genetic therapy.

It is easy to get caught up in the excitement (and, perhaps, hyperbole)[30] of gene therapy. Since the inception of clinical gene therapy in 1990, only a few hundred patients have been treated, and very few have been helped by these procedures. The entire genetic engineering enterprise yet may fizzle as a major advance in practical human medicine. On the other hand, it is perhaps just as likely that our great-grandchildren will look back with wonder at how their immediate ancestors managed to cope with the threat of debilitating, untreatable genetic disorders, much as we ourselves may marvel at the fortitude and toughness of our forebears in the days before public sanitation, anesthesia, surgery, antibiotics, and vaccines.

Ethical Quandaries from the Biotechnologies

The discoveries of the genetic sciences have uncovered a host of ethical issues that never before have been faced by our species.[31] These range from relatively crass (albeit important) social/economic questions of who should own the commercial rights to genetic discoveries, to refined quandaries about the biological future of humankind. Evolutionary science cannot provide definitive answers to such matters involving ethical principles. However, to approach the philosophical discussions scientifically unarmed is an unnecessary handicap. At the very least, responsible citizens should be aware of the profound challenges and opportunities to society afforded by the genetic revolution.

Rights to Genetic Property

Lest anyone doubt the potential impact of recombinant DNA technology, consider the ongoing scramble by private and public institutions to capitalize upon the new genetic technologies. Moral issues aside, such commercial hubris documents the widespread perception of tremendous economic potential from genetic engineering. Numerous biotechnology firms and government-funded groups (e.g., academic and federal research units) vie for slices of the biotechnology pie. Many diagnostic and therapeutic medical tools have emerged from biotechnological research on natural and engineered genes. The engineered drug with highest payoff to date (erythropoeitin, used to treat anemia) earns about $1.5 billion per year, but numerous other genetic biotherapeutics have comparable or greater economic potential. Clearly, the monetary stakes are high.[32] Many of the commercially viable products of biotechnology stem from the identification, characterization, and utilization of particular genes. Who has the right to profit from such discoveries? Who owns human genes?

In 1991, the National Institutes of Health (NIH) shocked many observers by filing an aggressive patent application for the first batch of about 1,250 ESTs (expressed sequence tags) partially sequenced from human tissues.[33] The attempt was stimulated by the Bayh-Dole Act of 1980, which encouraged government-funded research units to seek patent protection for their discoveries. The functions of these sequenced DNA fragments were unknown at the time, but the genes of which they are a part undoubtedly have importance to cells, and with further research and development someday might be used commercially. Nonetheless, the U.S. Patent and Trademark Office rejected these patent claims, and the NIH discontinued support for this large-scale cDNA sequencing project. The effort subsequently was taken over by profit and nonprofit organizations.

Other genetic patent attempts have been more successful. As of 1995, nearly 1,200 patents on human gene sequences had been granted worldwide, and currently about 450 formal applications are submitted per year.[34] For example, the U.S. companies Genentech and Kiren-Amgen own the patent rights respectively to "tissue plasminogen activator" (used to treat heart attack patients) and erythropoeitin; the Japanese firm Takeda is the legal proprietor of more than sixty genes; and the University of Washington and the University of California each own at least a dozen more. These organizations therefore possess the legal right to exclude others from making, using, and selling commercial products (e.g., a therapeutic drug or diagnostic procedure) from their genes, usually for a period of twenty years. Patents issued for human DNA range from PCR primers of diagnostic use to synthetic hybrids between interferon and interleukin genes that have therapeutic potential for treating allergies and chronic diseases such as arthritis. The most numerous patents (over a hundred) have been for genes with antitumor and antiviral functions, but other legally patented genes run the gamut from those encoding blood components and growth hormones to those that help to operate our pulmonary, vascular, neurological, immunological, reproductive, and digestive systems.

Patent laws exist to encourage invention through research and design. According to the Pharmaceutical Research and Manufacturers Association, it takes more than a decade and tens of millions of dollars, on average, before a new pharmaceutical product is ready for the market. Thus, it is doubtful that medically important applications of biotechnology will be developed aggressively without patent protection. On the other hand, granting patents indiscriminately every time a new DNA fragment is sequenced would impede further genetic discoveries. Patent law gradually has evolved a middle ground: successful patent applications must include discoveries about gene function that go beyond knowing simply the nucleotide sequence, and potential commercial applications must seem practicable.

Short of patenting, another approach currently used by biotechnology firms involves licensing agreements wherein, for a fee, gene sequences are provided to subscribers who have purchased exclusive or nonexclusive access to these data from which new technologies or products might be developed. For example, the rights to one newly discovered gene associated with obesity were sold in 1995 for $70 million.[35] Finally, some companies have found their best interests to be served by placing all of their newly obtained DNA sequences into the public domain. The motivation may be to facilitate cost-lessening collaborations with others, to promote company research and development by priming scientific advances based on the sequence data, or (from a cynical perspective) to undermine the research investments of competitor firms that may be stingier with their own DNA sequence information.

All such methodological approaches to the protection of intellectual property rights are merely pragmatic. What of the deeper ethical concerns? The following statement was issued by a group of nearly two hundred religious leaders at a press conference in Washington, D.C. on May 18, 1995:[36] "We, the undersigned religious leaders, oppose the patenting of human and animal life forms. We are disturbed by the U.S. Patent Office's recent decision to patent human body parts and several genetically engineered animals. We believe that humans and animals are creations of God, not humans, and as such should not be patented as human inventions." An accompanying press release clarified that the group's opposition extended to patents on DNA sequences.

This press conference was publicized widely as yet another clash between science (in this case, genetics) and religion. However, as Ronald Cole-Turner, a theologian, comments, such sentiments are far from universal even among religious representatives.[37] In a statement made in 1989, the United Church of Christ stated, "we welcome [genetic engineering's] development, pledging to support a climate of thoughtful reflection, public awareness, appropriate regulation and justice in distribution." Three years later, the United Methodist Church stated "Genetic techniques have enormous potential for enhancing creation and human life when they are applied to environmental, agricultural, and medical problems." With respect to patent issues themselves, the National Council of Churches (an affiliation of about thirty theistic denominations) already had taken a stand in 1986: "Scientists, investors and managers who provide the knowledge and capital necessary for biotechnological development and marketing deserve fair compensation for their ingenuity, work, and willingness to incur economic risks." According to Cole-Turner, "there is no distinctly religious ground for objecting to patenting of DNA." About genetic engineering, he concludes that "religious leaders who are both knowledgeable and humble are needed at the table of public discourse" and "when science and religion work together, there is at least the chance that we will be able to chart a responsible and sustainable future."

Ethical issues about intellectual property rights to DNA also arise in a somewhat different context. Alleles responsible for many human genetic disorders are prevalent or even confined to particular ethnic groups, or to small numbers of afflicted individuals.[38] In what is sometimes referred to as "genetic prospecting," biotechnology firms often focus on such groups as favorable natural sources from which to isolate and characterize disease genes, some of which later may prove to have commercial value.[39] For example, a patent recently was issued to the U.S. Department of Health and

Human Services for a human t-lymphotropic virus (HTLV-1) derived from the Hagahai people of Madang Province in Papua New Guinea. Should the patent holders be obliged to share with the Hagahai peoples whatever profits might derive from the unique genetic material which the latter originally donated? No, say some researchers, any more than a Pulitzer Prize winner should be forced to share her prize with the peoples she may have written about. Yes, say others, who point to the sorry history of colonialism and exploitation of third-world peoples by the industrialized West. In an extreme case, gene prospectors have been portrayed as new-age vampires drawing the blood of their victims for personal benefit.

There has been no consensus on such complicated issues. On the one hand, shouldn't bearers of disabling genes from which valuable products are derived be entitled to profit from their particular genetic endowment? On the other hand, to the extent that financial royalties to DNA donors diminish incentives for genetic research, the donors themselves and their families could stand to lose the most by inhibiting the development of diagnostic techniques and therapies for the diseased genes they bear. Also, royalties paid to DNA donors might contribute to the perception of individuals as commodities, one of the offensive hallmarks of earlier eras of colonialism.

Rights to Genetic Knowledge

Now that genetic predispositions for many straightforward metabolic disorders can be diagnosed routinely, who should have access to the clairvoyant information from this medical crystal ball? Suppose an individual is diagnosed as carrying the AIDS virus, or an allele for an early onset form of Alzheimer disease. Many parties in addition to the patient and her immediate family likely would have a keen and sometimes vested interest in this finding, including the woman's employer, her health and life insurance providers, and public health agencies. If the diagnosis became open knowledge, the woman might anticipate discrimination in the job or insurance market. On the other hand, if the test results were kept secret except to the woman or her family, a sort of reverse discrimination might arise. Suppose, for example, that the afflicted woman decided to purchase a large insurance policy under the terms of standard health and life expectancy. She or her family would be financial winners, and the company and its other policyholders the losers, since premiums inevitably must rise to cover such financial settlements. Traditionally, insurance has worked on the principle that policyholders pay premiums according to the risk they bring to the insurance fund, but until recently explicit genetic risk factors were unavailable. Who has the right to the new genetic knowledge?

Who decides the kinds and levels of genetic testing to be done? Employers and insurance companies no doubt would wish to have greater assurance that their employees and policyholders are genetically hardy, just as they now routinely require medical exams to assess the physical fitness of applicants. At the population level, many insurers would like to refine their actuarial tables to accommodate any gene-based mean differences among ethnic groups in health and longevity.

Such issues are not merely academic, and many countries currently are wrestling with legislative and procedural solutions. The following policies, current in February 1996 (situations often change rapidly), indicate the diversity of outcomes reached. The United Kingdom, Germany, Japan, Italy, Spain, and Portugal had no laws governing the use of genetic tests by insurance companies. In the Netherlands, life insurers legally could require a genetic test for any policy above $224,000. Belgium, Austria, and Norway had banned such requirements indefinitely. France had imposed a general moratorium on the use of genetic test data pending further study. In the United States, ten states prohibited genetic testing and the use of genetic data for health insurance purposes, but no such restrictions applied to life insurers.

In addition to the laws of governments and the policies of insurance companies, lobbying groups and nonprofit organizations have added their research and opinions to the debate. At one end of the continuum, the U.S. NIH/DOE Working Group on Ethical, Legal, and Social Implications of the Human Genome Project suggested that insurance companies be denied all access to genetic information, whereas the Genetics Interest Group in London (representing more than a hundred charities for people with genetic disorders) saw the use of genetic data by insurers as nearly inevitable, and urged that the procedures be regulated carefully by national legislation.[40] A variety of more specific suggestions has appeared as well. For example, insurance policies might be issued at standard rates, but if a patient succumbs to an "official" genetic disease the sum of benefits paid out would not exceed some ceiling, the remaining costs to be met by an industry-wide levy imposed on premiums for the listed genetic disorders. Insurance providers themselves are somewhat ambivalent on the issue of genetic data, with some viewing informational access as essential and others as an expensive and impractical base for refined actuarial calculations.

One central moral quandary relating to health coverage and life insurance is how to view the concepts of individual responsibility and accountability when a person's fate can be dictated by forces outside of his or her control, in this case by the genetic gods. It may be one thing to penalize a person financially for choosing an unhealthy lifestyle such as smoking or cave-diving, but quite another to compound life's obstacles for individuals who, through absolutely no fault of their own, already are burdened with a serious genetic disorder. Yet, as some insurance representatives are quick to point out, businesses are businesses and not social welfare institutions. They note that insurance firms charge higher rates for the elderly because life itself (rather than the insurance company) is unfair. If a genuine compassion for those with overt genetic disabilities is to be translated into substantive financial or other assistance, this may have to take place through avenues other than the economic thoroughfares of unfettered free enterprise.

Rights to Life and Reproduction

Particularly in Western societies, the right to life of a fetus, sometimes regardless of health risks to the mother, has been hotly debated in recent years. Such ethical issues for society fall outside the framework of science alone, yet biological information often is used (and abused) in the debate. What, if anything, can the evolutionary-genetic sciences add to the deliberations?

In the abortion wars, much attention has been devoted to the question of when life begins. According to Roman Catholicism, and to many fundamentalists, life is inaugurated at conception. Judaism and some Christian denominations preach that a human life is infused with a soul at birth. The Supreme Court ruled in *Roe v. Wade* that the state could not determine when life begins, and that government's obligation to protect an individual starts at the point when the fetus can exist outside the mother's womb.

All cells (including gametes and zygotes) arise from preexisting cells, such that in this sense there is an ineluctable continuity to life. At conception, two haploid cells wed in a marriage that will last until the death of the somatic individual to which they have given rise. The DNA blueprint in each zygotic nucleus is a particular combination of genes never before seen in the history of life. However, to suggest that each individual is unique and begins a new existence at conception is not to imply that early-stage embryos necessarily warrant insuperable rights.[41] Is it wise legislation to elevate the interests of small masses of developing cells in the uterus over those of a pregnant woman who bears the early embryo?[42] Our society seems unable to reach any consensus on this matter. Questions surrounding late-term abortions are even more troublesome.

The genetic sciences have revealed in considerable detail the mechanistic processes by which genes direct human development and operation. They also have revealed the many things that can go wrong. One common religious tenet with which scientific understanding appears incompatible is that the metabolic fates of developing embryos are governed by intelligent and caring supernatural forces. To the contrary, they are governed by natural gene-environment interactions that unfortunately can include such idiosyncratic molecular happenstances as whether an oxygen radical may have induced a particular mutation in the hypoxanthine-guanine phosphoribosyltransferase gene leading to Lesch-Nyhan syndrome, or whether a particular chromosomal nondisjunction event during meiosis has produced Down syndrome. The mere occurrence of such genetic conditions indicates we cannot trust omnipotent powers (or nature) to intervene against serious molecular-based disabilities. Should we then take the reins?

With the advent of genetic diagnostics, one potential "solution" is to influence human reproduction according to genetic criteria.

Selective abortion of genetically disabled fetuses is one such means, but so too are gametic screening, genetic counseling, and financial or other incentives for couples to refrain from perpetuating genes deemed deleterious. Any new eugenics must bear no resemblance to the horribly ill-guided movements at the turn of the last century, and the horror of eugenics under the Nazi regime should never be underemphasized or forgotten. But neither should the sad history of eugenics dissuade us from a careful consideration of humane opportunities presented by the current genetic technologies.

Many people already benefit immensely from the services of genetic counseling and molecular diagnostics in making reproductive decisions. Even the simplest bits of genetic information may circumvent decades of pain and suffering for entire families. The moral issues usually are easier when the genetic information is obtained prior to pregnancy, or earlier in embryogenesis rather than later, and when the projected genetic disabilities are undeniably horrible. Yet, experience indicates that no stance on reproduction or abortion will be morally agreeable to all, particularly when later stages of embryonic life are involved, and the anticipated genetic disabilities are milder. It would seem reasonable that the interests of the individuals most closely involved (the woman and her family) normally should take legal precedence over those of more distant parties.

Rights to Genetic Intervention

Since 1974, an august body of scientists and nonscientists known as the Recombinant DNA Advisory Committee (RAC) of the National Institutes of Health has faced the daunting challenge of anticipating and addressing ethical concerns pertaining to gene therapy and recombinant DNA practices. Formed at a time when genetic technologies first gave researchers the power to artificially splice together genes from different microbial species, the RAC ever since thoughtfully has policed the field of genetic engineering. More recently, another watchdog unit known as ELSI (Ethical, Legal, and Social Implications Branch of the Genome Office) was constituted and charged with addressing the many ethical and social policy questions that likely would arise from the Human Genome Project.[43]

Controversial though their recommendations sometimes have been, these committees have played an unprecedented role in the annals of science. More traditionally, scientific "progress" has taken place in relative ethical vacuums, societies being left to deal with the ramifications only after the fact. For better or worse, such was the case with the highly secretive Manhattan Project of the 1940s that provided the technological tools for nuclear energy and warfare. Even more remarkable (and laudatory) is the fact that geneticists themselves initiated, without outside pressure, both the self-policing regulatory policies and the extensive public discourse that have characterized the recent recombinant DNA revolution. This is not to say that the many ethical dilemmas related to biotechnology have been resolved.

At one end of the spectrum of public opinion is the view that any explicit manipulation of human biology is undesirable. Whether or not this philosophy has merit, it is operationally moot. For millennia, humans intentionally have altered our external and internal bodily environments in ways that influence health. In the medicinal arena, for example, metabolism-altering compounds have been employed for all manner of ailments, from gout to gonorrhea. Some of the first languages uttered by primitive peoples probably included words describing particular plants and animals perceived to be sources of desirable medicinal products. Modern pharmacology has continued this tradition, deriving many of its prophylactic and therapeutic drugs from natural sources. In this sense, the outcomes envisioned under many applications of gene therapy represent nothing fundamentally new. The production and delivery of metabolism-altering drugs will merely be shifted to engineered genes. So, many of the ethical questions that arise with conventional drug administration will reappear in the newer context of somatic gene therapy. What kinds of medical conditions should be treated, and how aggressively? Who gets treated? What is a "normal" as opposed to an "altered" metabolic state? What are the side effects? What misuses and abuses might arise? Thoughtful answers to such questions will require careful case-by-case appraisals.

Near the other end of the opinion spectrum is the dauntless proposal that human genetic engineering be extended to germ cell lineages. In contrast to somatic gene therapy that directly impacts only the individual recipients, germline genetic engineering in principle could alter genes in all future generations. Such a development truly would be revolutionary in the history of life on earth. Such applications involving the human germ line would require extensive ethical debate.[44]

The many ethical challenges prompted by the new genetic technologies are both complex and profound. In response, not only scientists, theologians, and lawmakers, but everyone must gather at the discussion table to consider rational, humanitarian courses of action. In such deliberations, perhaps the only mode of argument to be firmly censored—the only "wrong" approach—is that in which the moral authority of a god is asserted. As judged by the diversity of opinions held by responsible individuals on ethical matters pertaining to the human condition, any supernatural deity either has been strangely silent on such issues or else has conveyed vastly different messages to different listeners.

EIGHT

Meaning

Devoid of ultimate meaning, life means all the more.

Anonymous

"God," insisted the antelope, "is a runner, swift and free, who loves to leap and race with the wind."

"She is a great tree," murmured the willow, "a part of the world always growing and always giving . . ."

"She is a hunter," roared the lion. "God is gentle," chirped the robin. "He is powerful," growled the bear.

Douglas Wood, Old Turtle

The science of evolutionary genetics, in contrast to other belief systems, is descriptive rather than prescriptive of human affairs. Although science may illuminate why we behave and think as we do, and why our species may be inclined in particular ethical directions, it cannot tell us how we should act and think, or what our moral principles should be. Commandments of the form "thou shalt or shalt not . . ." reflect contextual ethical judgments whose social or biological origins, but not absolute merit, can be examined by objective scientific methods. This is not to suggest that theology and religion hold exclusive intellectual property rights to moral issues. By explaining how we have become who we are, science may help to weave an intellectually richer fabric of understanding for human morality.

The deities honored in most societies appear to have helped intelligent, group-oriented primates cope with life. Often made in man's image, these deities mirror humankind's behavioral and moral predilections, which evolved over millennia of biological and cultural evolution. By contrast, the genetic gods are material agents, outcomes of natural evolutionary processes that have shaped them and their organismal vessels, ourselves included.

The genetic gods may not warrant the kinds of worship and devotion traditionally reserved for supernatural deities—these DNA gods have no consciousness or sentient codes of conduct (notwithstanding a proclivity toward self-perpetuation), no reflective concerns about the consequences of their actions. Some would argue that they should not be referred to as gods at all: "[if] the word "God" is to be of any use, it should be taken to mean an interested God, a creator and lawgiver who has established not only the laws of nature and the universe but also standards of good and evil, some personality that is concerned with our actions, something in short that is appropriate to worship. This is the God that has mattered to men and women throughout history."[1]

Regardless of what they are called, genes are tangible entities, with profound influences on humanity. Indeed, over the last century, the genetic gods would seem to have wrestled from the supernatural gods considerable authority over human affairs. Does any room remain for a metaphysical god?

Pascal's Wager

The fact that the evolved mind of a biological creature such as man is capable of creating an image of a god says little about whether or not an ineffable reality of god actually exists. "It is incomprehensible that God should exist, and it is incomprehensible that He should not exist." So wrote Blaise Pascal in 1660.[2] Pascal was a scientist and philosopher best remembered for his theistic "wager." On the night of November 23, 1654, this thirty-one-year-old mathematician of Roman Catholic upbringing experienced a mystical religious experience (that he never fully disclosed) that changed his life profoundly and caused him to seek a more satisfactory dialectic between scientific "Reason" and matters of the "Heart." He came to believe that Reason alone cannot provide satisfying answers to questions of morality and ultimate meaning, but instead must be coupled to deistic convictions. To atheists and agnostics, he posed the following argument: If God does not exist, a person loses nothing by believing in him; but if God does exist, belief in him can bring eternal life. Thus, one should wager that God exists.

Pascal's assumption that an individual can choose whether or not to believe in God appears to clash philosophically with his own adherence to a stern form of Roman Catholicism (Jansenism) that accepted predestiny and rejected free will. This irony aside, at least two other questionable steps of logic underlie Pascal's wager. First, if God exists, Pascal assumed that only a belief in him can bring eternal salvation. However, by what logical or ethical rationale would God require human affirmation or damn nonbelievers? Pascal's reasoning merely suggests that a smart theological bet should be placed on any god that promises more, because if correct a person thereby wins a greater payoff. Second, Pascal's wager assumes that nothing is lost by mistaken belief in God. Given the empirical history of man's inhumanity to man, often in the name of a god, this assumption too is impugnable. As Pascal noted, "Men never do evil so completely and cheerfully as when they do it from religious conviction."[3]

Some of the most deeply held of human moral convictions involve the concept of theism itself. One of Fydor Dostoevsky's fictional characters notes that "what is strange, what is marvelous, is not that God really exists, the marvel is that such an idea, the idea of the necessity of God, could have entered the head of such a savage and vicious beast as man."[4] Yet it is now difficult to imagine human culture not possessed by faith in deities.

To the finite human mind, life can seem awesome, unintelligible, apparently engineered by a supreme force. Much of human culture depends on belief for order and stability, comfort and hope, self-righteousness and strength, revelational joy, and the promise of a glorious eternity. Personal empowerment through theism no doubt has enhanced the mean reproductive fitness of individuals. Furthermore, shared beliefs and their associated sacred rituals are powerful cohesive forces within groups, fostering collaborative efforts and unities of purpose that promote success in tribes or societies. E. O. Wilson suggests that "the predisposition to religious belief is the most complex and powerful force in the human mind and in all probability an ineradicable part of human nature."[5]

Whether or not a god exists, theistic beliefs do, and these appear to have been favored strongly in the biological and cultural evolution of our species. Mankind has produced approximately 100,000 religions. These have varied tremendously in the number and the nature of deities revered; one tribe's gods can be another's sacrilegious idols. Widespread monotheism is relatively new yet these religions have spread explosively. The demographic success of theism is indisputable. However, as evidenced by innumerable wars conducted and atrocities committed in the name of a god, religious victories often come at incalculable human cost. The net historical effect of theism on mankind's well-being remains open to debate.

The fundamentalist view that theism prevents the collapse of civilized societies from immoral behavior is opposed by views like that of Gore Vidal:[6] "The great unmentionable evil at the center of our culture is monotheism. From a barbaric Bronze Age text known as the Old Testament, three antihuman religions have evolved—Judaism, Christianity, and Islam. These are sky-god religions . . . The sky god is a jealous god . . . Those who would reject him must be converted or killed for their own good." The humanist Bertrand Russell also rejected theism as a civilizing force: he viewed all religions as abhorrent constructs motivated by fear— "fear of the mysterious, fear of defeat, fear of death. Fear is the parent of cruelty, and therefore it is no wonder if cruelty and religion have gone hand in hand."[7] Some would argue that humanitarianism, love, and compassion are among the widespread legacies of religions. However, so too are hatred and indifference. The beneficiaries of a religion tend to be its subscribers, whereas outsiders are more likely to suffer from those subscribers' beliefs. The suggestion that there exists a genetic predisposition for human belief systems does not prove that religion has not caused tremendous human suffering. If belief systems have profited our ancestors evolutionarily, in part it has been through the moral justification of violence against dissimilar cultures.

Even to some ardent theists, a god can be difficult to comprehend and is best accepted through faith. If human reason gave scant comfort in Pascal's day to concerns about an ultimate meaning for life, it would seem to provide even less solace in the wake of the discoveries of genetics and evolutionary biology. By all objective scientific evidence, our immediate biological fates, like those of other species, are influenced profoundly by genetic gods and other natural forces. The genes themselves have been molded by amoral and unconscious evolutionary processes such as mutation, recombination, and natural selection. Thus, "the evolutionary process is purposeless and uncaring . . . Our modern understanding of evolution implies . . . that ultimate meaning in life is nonexistent."[8]

Charles Darwin wrestled with the theological questions that his own discoveries raised. Darwin had been brought up in a Victorian English society dominated by the religious orthodoxy of the Anglican church, and although he appears never to have been particularly captivated by religious convictions, neither did he shy away from many of the Church of England's general philosophical tenets. In his youth, Darwin had no difficulty, for example, in reconciling his developing interest in biology with a "natural theology" popular at the time that interpreted nature's apparent design as abundant proof of a creator.

In later years, however, Darwin's understanding of natural selection made it increasingly difficult for him to reconcile his scientific knowledge of nature's operations with the Church's insistence on the Old Testament's literal authority on matters of history, or on its acceptance of supernatural miracles, or on its convictions about moral truths stemming from divine revelation. Darwin increasingly came to view the evolutionary record as the correct scribe of biological history, of life's operations as being understandable through the study of natural mechanisms, and of moral certitudes representing human-specific systems of "belief allied to instinct."[9] Darwin reached these conclusions almost reluctantly through his scientific findings. His new ideas created travails in his loving marriage to his devout wife and in his attempts to cope with the death of his ten-year-old daughter. Darwin eventually became an atheist in the sense that he could no longer subscribe to a belief in the direct actions of God, but he never wore this conviction on his sleeve as a point of pride or satisfaction.

Evolutionary biology and genetics are only the latest of the sciences whose findings appear to challenge traditional theological tenets. Discoveries in geology and astronomy concerning the physical age and structure of the planet and of the universe have contradicted many religious beliefs. Albert Einstein once stated that he believed in a god who "reveals himself in the orderly harmony of what exists, not in a god who concerns himself with fates and actions of human beings."[10] If a god's hand is manifest in the day-to-day operations of the universe, Einstein believed it would be via the laws of science.[11]

What the evolutionary-genetic sciences point to most clearly is the important influence of genes over many human affairs that were thought to be under the purview of supernatural deities. The genes exercise these powers not in a vacuum, but rather in intimate collaboration with physical and social environmental conditions to which we are exposed during our development. The genetic gods and their protein angels interact elaborately with one another, and with environmental factors ranging from intra-cellular to macro-ecological. Many environmental conditions themselves, notably human cultures, reflect extended influences of the genes in this and prior human generations. The outcome is an individual person, unique from all others who have come before or ever will follow.

Yet the genetic gods have evolved according to understandable, mechanistic biological processes such as mutation and DNA repair, recombination, Mendelian transmission, and Darwinian selection. Only natural selection comes close to omnipotence, but even here no intelligence, foresight, ultimate purpose, or morality are involved. Natural selection is merely an amoral force, as inevitable and uncaring as gravity.

Evolutionary genetics confronts us with a grand version of the Tiresias dilemma: Is it but sorrow to be wise when wisdom profits not? What is to be gained by an awareness of genetic operations and evolutionary processes when such knowledge challenges our faith in a loving and interventionist god? Unfortunately, science provides no assurance that the knowledge it uncovers will better human existence. We might be far more content believing that the universe is earth-centered, or that humans are the focus of a god's affection. Scientific discoveries may be lamented, but it is difficult to imagine how we might ignore them.

Is it possible that we may gain some philosophical insight from science? I prefer the optimistic stance, and not merely because we have no other options. Human existence is not meaningless just because life has arisen from natural processes and ends quickly for the individual. Human life can be abundant, robust, and immensely satisfying, although also nasty, brutish, and short.

An Evolutionary-Genetic Wager

In the tradition of Pascal, perhaps a new wager can be posed. If mortal life is all that exists for individuals, we lose nothing by seeking to make that life as meaningful and rewarding as possible. But if eternal life exists, we have lost nothing by seeking a fulfilling existence here on earth. Thus, one might wager on the richness of life here and now.

Like Pascal's original bet, this evolutionary-genetic wager involves some questionable assumptions. It assumes that nothing is to be lost by a mistaken belief in the absence of a god or of an eternal existence for the individual's soul. Many religions posit that only through complete faith can final redemption be attained. A far less severe philosophy holds that no deity would damn a soul for a lack of faith on matters unresolved to an open and reasonable, yet finite, human mind. Furthermore, some philosophers claim that, as a justification for ethical behavior, absolute faith is essential to society, regardless of its reality.

A second assumption of the evolutionary-genetic wager is that humans can choose to focus on the enhancement of meaning in immediate life, rather than in the hereafter. Findings from the biological and social sciences are ambivalent on this issue. Theism is a coping device from which many people derive great comfort and fulfillment. Recent findings suggest that this religious proclivity is influenced by genes, both as an innate and encultured part of human nature. On the other hand, if belief systems help humans address life's challenges of survival and reproduction, little additional philosophical incentive is required.

Unfortunately, history documents that the pursuit of individual agendas, when coupled with the human tendency to invent personal justifications for moral authority, has promoted innumerable religious wars and persecutions. In any conciliation of faith and science, the practical challenge is to identify and promote systems of understanding that contribute to personal enrichment and minimize harm to others.

Toward Enhanced Proximate Meaning

The emergence of *Homo sapiens* under natural evolutionary processes can be interpreted as even more miraculous and awe-inspiring than human creation by a god. It took more than four billion years of biological evolution to set the stage for human life. Even after the appearance of higher primates, thousands of millennia of proto-human evolution transpired before increasing mental capacities permitted knowledge to be transferred from one generation to the next. Throughout this long process, there was no predetermined script, no conscious direction, no inevitability of outcome. We might never have appeared. Now here, we may not survive for long. Insights from the evolutionary sciences challenge us to cherish human existence.[12]

In *The Mountain People,* cultural anthropologist Colin Turnbull describes the difficult lives of the Ik villagers in the barren highlands of east-central Africa. Previously, the Ik apparently had been a loosely organized society of prosperous hunters and gatherers with a rich culture not unlike those of other tribes in the area. Now, driven to the very edge of starvation, the Ik society in less than two generations had become little more than an unconnected assemblage of individuals pursuing his or her personal survival, by all appearances without family structure or cooperative sociality. In a sort of collective autism, the Ik now sought only to avoid starvation, and were forced to abandon what most of us consider to be the elements of a meaningful life.[13]

The Ik remind us that life on earth can become hellish, and that no omnipotent safety net can be assumed. Yet life can go on even in times of extreme deprivation. Thus, any serious philosophical discussion of human affairs must include concerns about the quality of life on earth, and not life's mere quantity or continuance alone.

If humans are to achieve any thoughtful, active influence over the thoughtless evolutionary-genetic processes that otherwise govern our fates, three routes are available: adjustment of genes, adjustment of environments, or both.[14] We can contemplate the conscious modification of our genes to better suit our physical and cultural environments, or the modification of our physical and social environments to better suit our genes. An assumption underlying either approach is that the prospects for human happiness and satisfaction in this life are enhanced by an appropriate match, or fit, between our evolved genetic endowments and the environments to which they are exposed.[15]

Notwithstanding the recent developments in genetic engineering, direct manipulation of the human genome is not a realistic means to alter the general human condition now. DNA technologies are too rudimentary, the costs too high, the logistic difficulties far too great, and the ethical ramifications not yet adequately considered. On the other hand, environmental engineering is well within our grasp, and indeed has been practiced widely by human cultures for hundreds of thousands of years, beginning with such achievements as the invention of tools and the domestication of fire.

Tool use is not unique to humans,[16] but it has been elaborated in our species to unprecedented lengths. From the simple stone tools used and shaped by our Australopithecine ancestors more than a million years ago to today's airliners and nuclear power plants, we have invented increasingly refined instruments that for better and sometimes for worse have altered profoundly our environment. For example, one important early development was the domestication of fire. Manmade hearths have been discovered in Europe and Asia that date to the Middle Pleistocene, some five hundred thousand years ago.[17] With no change in our genes, the controlled use of fire immediately provided our ancestors with a powerful new defense against predators, a source of light to open the night, warmth to withstand colder climates and colonize higher latitudes, an energy source for the construction of refined tools, and a means of cooking food from many otherwise inedible plants and animals. Equally profound changes were initiated about ten thousand years ago when humans domesticated animals and plants as sources of food, labor, and fiber.

Given the power of the human species over environmental conditions, what seems remarkable is how little attention has been devoted to thoughtful collective deliberations on the kinds of physical and social environments we might wish to promote for ourselves. To emphasize this point, consider the current issue of human overpopulation and the host of associated global changes that threaten the continuance of our species and many others. Informed scientists for decades have voiced fears about the human population bomb,[18] and in 1993 the most prestigious scientific academies of fifty-eight countries issued the following chilling conclusion:[19]

> It took hundreds of thousands of years for our species to reach a population level of 10 million, only 10,000 years ago. This number grew to 100 million people about 2,000 years ago and to 2.5 billion by 1950. Within less than the span of a single lifetime, it has more than doubled to 5.5 billion in 1993 . . . [Providing] fertility declines to no lower than 2.4 children per woman [the] global population would grow to 19 billion by the year 2100, and to 28 billion by 2150 . . . If current predictions of population growth prove accurate and patterns of human activity on the planet remain unchanged, science and technology may not be able to prevent irreversible degradation of the natural environment and continued poverty for much of the world . . . Humanity is approaching a crisis point with respect to the interlocking issues of population, environment, and development . . . In our judgment, humanity's ability to deal successfully with its social, economic, and environmental problems will require the achievement of zero population growth within the lifetime of our children.

Relatively few people are cognizant of the dire ramifications of the human population explosion. The Catholic Church promotes unrestrained human reproduction through its official positions on sex and abortion. The leaders of most countries, including the United States, generally have failed to acknowledge that a demographic crisis exists, let alone have they sought to identify humanitarian solutions to the fundamental problem of human overpopulation.[20]

Such laissez-faire attitudes about environmental issues are hardly new. When North America was colonized by Europeans five hundred years ago, little thought was given or action taken to circumscribe human impacts on the environment. Nearly the entire continent was deforested,[21] drained of wetlands, stripped of minerals and groundwater, polluted, dammed, paved,[22] plowed, and cultivated. Huge populations of buffalo and other wildlife were slaughtered, and native peoples massacred, all to accomodate short-term capitalistic appetites. Today we continue many of these activities even as we begin to appreciate the enormity of the cost. Current problems include massive urbanization with associated crime and poverty, global climatic change through emission of greenhouse gases, the destruction of habitat leading to the collapse of ecosystems and to increased rates of extinction, extensive toxic pollution, and an approaching exhaustion of fossil fuels, underground aquifers, and other nonrenewable resources. In the past, science and technology both facilitated and compensated for these crises. However, in the face of the burgeoning human population, future technological developments alone will not be enough.

Assuming that our planet's biosphere and its human occupants somehow survive the environmental crises of the twenty-first century, perhaps a time can be envisioned when societies would devote far more attention and conscious effort toward cultivating the kinds of social, physical, and biotic environments that might better suit human biology. Some issues to consider would be the composition and distribution of landscapes,[23] the sizes and organizations of towns or cities, opportunities for creating biodiverse environments,[24] and the sociopolitical and religious climates that might permit the greatest flowering of human possibilities. We all need to assess what environmental conditions promote a meaningful existence.

Toward a Reconciliation

In the current social and political climates influenced by fundamentalist religious movements, it is sometimes hard to entertain great hope for constructive accommodation between the biological sciences and religion. Fundamentalism demands absolute faith. Science espouses open-minded yet critical inquiry. Fundamentalism has no room for alternative explanations. Science welcomes alternative testable hypotheses for evaluation against evidence. Fundamentalism is antihistorical and anticontextual. Evolutionary biology and genetics deal with historical and contextual outcomes.

While the views of fundamentalist movements do not fairly

represent the whole religious spectrum, conundrums and ironies abound also in the broader clash between science and religion. Our global society capitalizes on the knowledge that science provides, yet receives inspiration and guidance from the religious beliefs that that knowledge erodes. The natural sciences have revealed so much about human origins and human nature, yet most religions have failed to accept or even to consider its findings. Intolerance of science by religious groups is relatively new in the course of human history. Many religions have integrated new knowledge about the earth to better explain the workings of the universe. Old cults usually left room for new gods in their pantheons. Greek philosophers such as Plato and Socrates were inspired to visions of a glorious universe through their studies of science and metaphysics, and later societies sometimes viewed them as being closer to God for their insights. In medieval times, St. Thomas Aquinas in his *Summa Theologiae* wrote that a grand synthesis of Christianity and the natural sciences could be achieved because the worlds revealed by sense and by faith would converge.[25]

The Koran stresses that thoughtful intelligence is needed to decipher messages of God delivered through the natural world, and it avidly encourages Muslims to examine their surroundings with curiosity and attentiveness. In the ninth century, a new movement within Islam became dedicated to the proposition that one should live in accordance with the laws of the cosmos, which could be discerned through the study of astronomy, medicine, mathematics, and the natural sciences. This led to a cultural florescence within the Abbasis empire. The Arab Faylasufs who led this movement interpreted rationalism as the most advanced form of religion. Rationality, they thought, could only refine the concept of God and free it from anthropocentrism and superstition.

Although perceptions of "rationality" may vary, even the more philosophically polarized of humans often are closer on general ethical persuasions than otherwise might be supposed. All members of our species share elements of genetic heritage that influence behaviors and moral views. We may argue vehemently about whether or not to ban abortions, but the issues themselves are deemed important because *Homo sapiens* is predisposed by genetic evolution to be concerned about individual and group survival.

We may pledge allegiance to the John Birch Society, the Libertar-
ian Party, or the Boy Scouts of America, but all reflect our species'
innate proclivity to form social alliances. Depending almost entirely on upbringing and cultural circumstance, we may dislike
Israelis or Palestinians, Protestants or Catholics, Communists or
capitalists, scientists or fundamentalists, the Yankees or the Dodg-
ers, but all of these reflect shared xenophobic tendencies (the other
side of group allegiances). Atheists, agnostics, and theists have
reached different conclusions about a supernatural god, but the fact
that the deliberations have taken place indicates that all camps have
contemplated and attach significance to such questions of ultimate
concern.

Despite outward appearances of otherworldliness, theism above
all is highly pragmatic. It is far more important to most individuals
that a particular idea of a god works for them, than for the idea to
be sound logically or scientifically. Conversely, for all of its apparent worldliness, basic science is ideally concerned only that conclusions be logical and objectively verified, not that they yield
utilitarian outcomes. Much of the current debate between science
and religion on "ways of knowing" reflects a failure by many
scientists to appreciate that a correct scientific explanation is not
necessarily right in any utilitarian sense, and a failure by the religious to appreciate that utilitarian rightness implies nothing about
scientific truth. Because science seeks truth regardless of utility, and
religion seeks utility regardless of truth, neither can be relied upon
entirely to extricate us from life's difficulties.

We are a philosophically changeable species with strong and
innate proclivities to exercise critical thought and at the same time
to seek inspiration and meaning through faith. We find ourselves
now in a crowded, crisis-riddled world where the evolutionary
legacies of scientific rationalism and religious pragmatism come
face to face as alternative modes of problem solving. It is unlikely
that we can divorce ourselves from either of these two sides of
human nature. Can a workable marriage between science and
religion nonetheless be achieved that will allow our species to face
the future with intelligence and inspired hope?

On this question, at least some scientists and religious leaders are
optimistic. Cardinal Joseph Bernardin recently stated,[26] "The hu-

man potential for creativity is being fulfilled in our day in many ways. Human learning in general, the work of the laboratory and technological advances in particular are not to be feared, but rather prized and celebrated as both a gift and a responsibility." Lines of communication between science and religion have opened on many fronts. For example, in recent years the U.S. National Academy of Sciences has helped the National Conference of Catholic Bishops' Committee on Science and Human Values develop an active dialogue with the scientific community on topics such as human population and the environment, genetic testing, genetic screening, and how to deal with death.[27] Throughout his papacy, Pope John Paul II has sought to reconcile science and faith, and in a statement issued in the fall of 1996 he declared that "new knowledge leads us to recognize in the theory of evolution more than a hypothesis . . . The convergence, neither sought nor induced, of results of work done independently one from the other, constitutes in itself a significant argument in favor of this theory."[28]

On the environmental front, several religious groups in the United States[29] recently organized the "National Religious Partnership for the Environment" (NRPE) with the goal of nurturing a more appreciative view of nature. This group is not composed primarily of conservation biologists, who sometimes approach their role with a religious fervor, but rather of religious practitioners who wish for a more reverent approach to the natural world. As one Lutheran minister stated, "The church can address the deeper issues of our whole relationship with creation, so that the changes we're able to make are not simply technological, but grow out of a deeper relationship and rootedness in nature."

Like the Arab Faylasufs of the last millennium, perhaps we can envision a new enlightenment or conceptual renaissance that incorporates rational understanding, in this case from evolutionary biology and genetics, into a deeper appreciation of the world. The genetic gods' influence over human fates is profound, but so too was the presumed dominion of the sky gods in many traditional religions. Both genetic and theistic determinism pose questions about the concept of free will, but the empirical truth remains that humans can and have created many different physical and cultural existences. The deeper challenge is to incorporate science's objective understandings of nature into broader philosophical frameworks and responsible modes of action that may help us find satisfying lives.

The fields of evolutionary biology and genetics have given unprecedented insight into biological mechanisms, but have contradicted many of the traditional tenets of theology and religion. Toward the end of his career in physics, Albert Einstein concluded: "In their struggle for the ethical good, teachers of religion must have the stature to give up the doctrine of a personal God; that is, give up the source of fear and hope which in the past placed such vast power in the hands of priests. In their labors they will have to avail themselves of those forces which are capable of cultivating the Good, the True, and the Beautiful in humanity itself. This is, to be sure, a more difficult but an incomparably more worthy task."

Einstein stated in interviews that a childlike openness and curiosity contributed to his success as a physicist. In that spirit, the following sentiment from the children's book quoted in the epigraph seems an appropriate conclusion to this chapter. Recall that a verbal argument had been going on between antelopes, trees, lions, robins, and bears over the nature of God. Eventually, a new plea was echoed. It seemed to come from the stones and rocks, from the mountain, the ocean, and the stars:

"Please, STOP . . ."
And after a long, lonesome and scary time . . . the people listened, and began to hear . . . and to see God in one another . . . and in the beauty of all the Earth. (Douglas Wood, *Old Turtle*)

Epilogue

To provide symmetry with the prologue, I will close this book with reference to another film classic of science fiction: *2001: A Space Odyssey*. The protagonist is HAL, a supercomputer with advanced machine intelligence rivaling that of the human brain, including programmed self-consciousness and a capacity for emotional expression. HAL and five human astronauts constitute the crew of the Discovery 1 spacecraft, whose assignment is to track down evidence pointing to extraterrestrial intelligence on the planet Jupiter. During the long journey, HAL becomes concerned that fallible humans might jeopardize the mission, and it endeavors to wrest full control of the spacecraft by killing its human compatriots. However, one astronaut escapes, and retaliates by disabling the circuit boards of HAL's higher thought functions. By aspiring to subordinate his human creators, HAL precipitated his own demise.

As we rapidly approach the year 2001, we find that our genes have inadvertently bestowed upon us the intellectual and technological capacity to contemplate challenges to the authority of the genetic gods for the first time in the history of life on earth. Capabilities both for genetic engineering and for extensive environmental alteration have emerged rapidly. How should such powers be exercised? In what images do we wish to shape ourselves and our environment in the continuing evolutionary odyssey? Should we endeavor to assume authority over our genetic gods, perhaps in directions that they themselves have predisposed us to think morally proper? Will genetic and environmental tampering improve the human condition? I hope this book will have stimulated thought about such questions. In the words of HAL, "This mission is too important . . . to jeopardize."

218

NOTES

GLOSSARY

INDEX

Notes

Preface

1. Many humanists in this century have considered the relevance of the biological sciences to theology and religion. Some recent books that provide introductions to this literature include: I. G. Barbour, *Religion in an Age of Science* (San Francisco: Harper & Row, 1990); M. Bradie, *The Secret Chain: Evolution and Ethics* (Albany: State University of New York Press, 1994); L. B. Gilkey, *Nature, Reality, and the Sacred: The Nexus of Science and Religion* (Minneapolis: Fortress Press, 1993); P. Hefner, *The Human Factor: Evolution, Culture, and Religion* (Minneapolis: Fortress Press, 1993); P. Kitcher, *Vaulting Ambition: Sociobiology and the Quest for Human Nature* (Cambridge, Mass.: MIT Press, 1985); M. H. Nitecki and D. V. Nitecki, eds., *Evolutionary Ethics* (Albany: State University of New York Press, 1993); R. J. Richards, *Darwin and the Emergence of Evolutionary Theories of Mind and Behavior* (Chicago: University of Chicago Press, 1987).

1. The Doctrines of Biological Science

1. G. C. Williams, *Natural Selection: Domains, Levels, and Challenges* (New York: Oxford University Press, 1992).

2. In the 1960s, a "God is dead" movement swept across the United States, fueled by the perception that a living god would not so neglect the plight of humans. See the philosophical discussions in J. B. Metz, ed., *Is God Dead?* (New York: Paulist Press, 1966). Some have thought instead that a well-intentioned god is overmatched. As expressed in subway graffiti, "God is alive and well but considering a less ambitious project."

3. Many other scientists have expressed similar sentiments. Mathematical physicist Paul Davies, an expert on black holes, time warps, and quantum mechanics, has written several books [such as *God and the New Physics* (New York: Simon & Schuster, 1983) and *The Mind of God* (New York: Simon & Schuster, 1993)] that infuse science into theology and spiritualism. A basic thesis of these treatments is that science offers a surer path to a god than does religion.

4. Even in this most difficult of arenas for study, scientific understanding is

221

growing rapidly. For example, there is considerable experimental evidence that discrete components of language recognition and use are localized to particular regions of the brain. Further mechanistic dissection of neural pathway operations may lead to a far deeper understanding of why brain damage from strokes or other sources can be so selective with respect to the loss of specific conceptual categories and classes of semantic memory. For an introduction to the brain's mechanistic operations, see: A. Caramazza, "The Brain's Dictionary," *Nature* 380 (1996): 485–486; N. C. Andreasen, "Linking Mind and Brain in the Study of Mental Illnesses: A Project for a Scientific Psychopathology," *Science* 275 (1997): 1586–1593; A. G. Cairns-Smith, *Evolving the Mind* (Cambridge: Cambridge University Press, 1996).

5. G. J. Mendel, *Versuche ueber Pflanzenhybriden,* Verhandlungen des Naturforschenden Vereins (Bruenn) 4 (1865): 3–47.

6. *On the Origin of Species* is a general work of natural history, and contains no direct reference to human evolution beyond a single sentence of classic understatement in the concluding chapter: "Much light will be thrown on the origin of man and his history." Later, Darwin did discuss human biology and evolution at greater length, for example in *The Descent of Man* (1871) and in *The Expression of the Emotions in Man and Animals* (1872).

7. Pre-Darwinian naturalists often interpreted the beauties and perfections of their biotic surroundings as compelling evidence for a creator's beneficence. For example, a travelogue diary detailing the expedition of one famous naturalist through the southeastern United States in the late 1700s is sprinkled generously with such statements as: "This world, as a glorious apartment of the boundless palace of the sovereign Creator, is furnished with an infinite variety of animated scenes, inexpressibly beautiful and pleasing, equally free to the inspection and enjoyment of all his creatures" (*Travels of William Bartram,* New York: Dover, 1955).

8. D. L. Hull, "Universal Darwinism," *Nature* 377 (1995): 494.

9. The Gaia hypothesis, proposed by James Lovelock in 1988 (*The Ages of Gaia,* New York: Norton), is an extreme vision of how natural selection might operate on a global scale. The notion is that natural selection has shaped the interactions of species and communities within the biosphere in such a way as to homeostatically adjust geophysiological processes, thereby maintaining earth's conditions (such as concentrations of atmospheric gases) at states advantageous for life. This idealistic scenario has more emotive appeal than critical scientific support.

10. Earlier in this century, much attention was devoted to the possibility that natural selection might operate directly to forge adaptations that serve the collective good of a species. These "group selection" arguments reached their apogee in a book published in 1962 by V. C. Wynne-Edwards: *Animal Dispersion in Relation to Social Behavior* (Edinburgh: Oliver & Boyd). Such views now are discredited by most evolutionary biologists.

11. A recently reported example involves a species-rich assemblage of columbine plants, in which a key innovation in the morphology of nectar spurs (in response to pollinator availability) concomitantly reduced gene flow between populations, speeding up speciation. See S. A. Hodges and M. L. Arnold, "Spurring Plant Diversification: Are Floral Nectar Spurs a Key Innovation?" *Proc. Royal Soc. London* B 262 (1995): 343–348.

12. Some disputed evidence remains for *directed mutations* in bacterial and yeast evolution, whereby mutations that confer a selective advantage to altered environments may tend to arise preferentially when needed. For a recent review, see P. D. Sniegowski and R. E. Lenski, "Mutation and Adaptation: The Directed Mutation Controversy in Evolutionary Perspective," *Annu. Rev. Ecol. Syst.* 26 (1995): 553–578.

13. R. Dawkins, *The Blind Watchmaker* (New York: Norton, 1986).

14. Darwin, *On the Origin of Species*.

15. Phylogeny is the evolutionary history of a group or lineage. As noted by the famous paleontologist George Gaylord Simpson in 1945, "The stream of heredity makes phylogeny; in a sense, it is phylogeny."

16. R. M. Nesse and G. C. Williams, *Why We Get Sick* (New York: Random House, 1994).

17. See, for example, ibid.; G. Estabrooks, *Man, the Mechanical Misfit* (New York: Macmillan Co., 1941); E. Morgan, *The Scars of Evolution* (New York: Oxford University Press, 1994).

18. This movie analogy was developed eloquently in: S. J. Gould, *Wonderful Life: The Burgess Shale and the Nature of History* (New York: Norton, 1989).

19. In 1978, the evolutionary biologist E. O. Wilson was among the first of prominent scientists to dare an explicit exposition on the biological basis of human nature (*On Human Nature*, Cambridge, Mass.: Harvard University Press); an accomplishment for which he received the 1979 Pulitzer Prize for general nonfiction. This volume was an extension of the concluding chapter from his 1975 book (*Sociobiology: The New Synthesis*, Cambridge, Mass.: Harvard University Press) that had dealt with social behavior in nonhuman species. In his recent autobiography, *Naturalist*, (Washington D.C.: Island Press, 1994), Wilson describes the heated controversy and deep hostility, particularly from Marxist scholars, that met these early attempts to apply evolutionary principles to human behavior.

20. The recent poll was reported in: E. J. Larson and L. Witham, "Scientists are Still Keeping the Faith," *Nature* 386 (1996): 435–436.

21. Many other evolutionary biologists have expressed similar opinions. For example, Theodosius Dobzhansky wrote extensively on the philosophical ramifications of biology in his book, *The Biology of Ultimate Concern* (New York: New American Library, 1967).

22. Issued in 1988, this statement is one of many by Pope John Paul II affirming an openness to science.

2. Geneses

1. A notable exception is the Indian religious system of Jain, which is explicitly noncreationist. Adherents to Jainism point out the foolishness of asserting a creationist god by noting questions such as the following: From whence could a god himself come? How could a nonmaterial being make a material world, and why would a god wish to do so? Why would a perfect being will the creation of something less than perfect? Why would a creator kill or cause death among these creations? Instead, Jainism views the world as without beginning or end, though nonetheless divided into earth, heaven, and hell.

2. Of course, the question "Who made God?" remains open under such accounts.

3. This is an example of an "earth-diver" story, one of the most common types of creation myths, particularly in Native American cultures. A supreme being typically sends a creature such as a turtle or crawfish into primal waters to retrieve clay or pebbles from which the earth or its living inhabitants are created.

4. As retold in A. Eliot, *The Universal Myths: Heroes, Gods, Tricksters and Others* (New York: New American Library, 1990).

5. In an Aztec creation story, the Gods Quetzalcoatl and Tezcatlipoca pulled the earth goddess Coatlicue from the heavens and ripped her asunder to form the earth and sky. Coatlicue's hair became plants, her eyes and mouth became caves and springs, and other body parts became mountains and valleys. Coatlicue's anger at this treatment explains why she demands sacrifices of human heart and blood.

According to a Boshongo myth from the Bantu peoples of Central Africa, in the beginning the great Bumba vomited up the sun, the moon and the stars, and various animals such as the crocodile, tortoise, heron, and human, which soon begat variations of their types. In an Eskimo creation story from the Chukchee peoples in northeastern Siberia, Creator-Raven defecated and urinated as he flew, and his droppings became the original mountains, rivers, and lakes. In a related creation myth of the Kodiac Island Eskimos, the first woman (created by Raven) urinated and spit to make the oceans and bodies of fresh water.

6. In the United States, demands often arise for either a removal of evolutionary topics from science curricula, or equal time for creationism in the science classroom. Most scientists argue that the appropriate place for creation myths in school curricula is in religion or history courses. For a typical courtroom case, and one judge's enlightened decision, see "McLean versus the Arkansas Board of Education," *Science* 215 (1982): 934–943.

7. For a comprehensive review of geological time assessments based on decay rates of long-lived naturally occurring radioactive isotopes (and from other lines of

evidence), see G. B. Dalrymple, *The Age of the Earth* (Stanford, Calif.: Stanford University Press, 1991).

8. Stromatolites are columnar or mound-shaped mineral deposits accreted from matlike communities of microscopic organisms. Living stromatolite communities still exist today in several places, including the coast of Western Australia.

9. Concrete evidence that life exists or did exist on other planets such as Mars would have a major scientific impact, because it would strongly suggest that life can arise commonly under suitable natural conditions. A controversial report of potential life on Mars is D. S. McKay et al., "Search for Past Life on Mars: Possible Relic Biogenic Activity in Martian Meteorite ALH84001," *Science* 273 (1996): 924–930. See also M. Grady, I. Wright, and C. Pillinger, "Opening a Martian Can of Worms?" *Nature* 382 (1996): 575–576; C. F. Chyba, "Life Beyond Mars," *Nature* 382 (1996): 576–577; and note 11.

10. S. L. Miller, "A Production of Amino Acids under Possible Primitive Earth Conditions," *Science* 117 (1953): 528–529. See also S. L. Miller and L. E. Orgel, *The Origins of Life on Earth* (Englewood Cliffs, N.J.: Prentice-Hall, 1974). For a detailed but readable introduction to the early fossil record of life, as well as to evidence on the prebiotic synthesis of organic compounds, see the relevant chapters in J. W. Schopf, ed., *Major Events in the History of Life* (Boston, Mass.: Jones & Bartlett, 1992).

11. Organic compounds also are thought to be present in the atmospheres of some other planets, as well as in "carbonaceous" meteorites that occasionally strike the earth. For example, one such meteorite from Mars, discovered in Antarctica, contained apparently indigenous polycyclic aromatic hydrocarbons suggestive of biogenic processes. Such observations also have raised an alternative possibility that the earth early in its history was "seeded" by life from elsewhere, a view notably championed by F. Crick in *Life Itself* (London: Macdonald & Co., 1981). Crick is one of the codiscoverers of DNA's structure. A related scientific possibility known as the panspermia hypothesis proposes that life is widespread and has moved around the galaxy by space-borne microorganisms. See the following and references therein: P. Parsons, "Dusting Off Panspermia," *Nature* 383 (1996): 221–222. If the earth was biologically seeded, questions concerning life's origin are merely pushed back to another time and place.

12. These are among the reasons why any organic molecules spontaneously arising in the modern oxygen-rich and biotic-rich world would have little chance of persisting long enough to instigate continuing geneses for life. Note also that almost all oxygen in the present atmosphere is the result of biological activity (photosynthesis) postdating life's origin in the earth's original reducing environment.

13. RNA, or ribonucleic acid, is a very close molecular relative of DNA. RNAs come in several distinct varieties that generally play a role in converting the coding information in DNA into proteins.

14. See, for example, T. R. Cech, "RNA as an Enzyme," *Sci. Amer.* 255 (1986): 64–75.

15. S. A. Kauffman, *The Origins of Order* (New York: Oxford University Press, 1993).

16. An increased opportunity for life's origin does not necessarily mean increased likelihood of the event. With the current sample size of one (life on earth), we have no way of stating empirically the likelihood of life's origins under "suitable" conditions here or elsewhere in the universe.

17. For comprehensive reviews of human evolution from morphological as well as genetic and other perspectives, see S. Jones, R. Martin, and D. Pilbeam, eds., *The Cambridge Encyclopedia of Human Evolution* (Cambridge: Cambridge University Press, 1992); and P. Mellors and C. Stringer, eds., *The Human Revolution* (Edinburgh: University Press, 1988). For a popular account of mankind's place in the biological world, see R. Lewin, *Human Evolution,* 3rd ed. (Boston, Mass.: Blackwell, 1993).

18. Interested readers might consult J. C. Avise, *Molecular Markers, Natural History and Evolution* (New York: Chapman & Hall, 1994).

19. Details of the numerous molecular genetic assays are far beyond the current treatment, but a good introduction is in D. M. Hillis, C. Moritz, and B. K. Mable, eds., *Molecular Systematics,* 2nd ed. (Sunderland, Mass.: Sinauer, 1996).

20. The genetic distance between humans and chimpanzees as estimated by protein electrophoresis is approximately 0.50, meaning that at roughly 50 percent of assayed genes, no electrophoretically detectable mutational differences distinguish a human from a chimpanzee.

21. For example, the following proteins from humans and chimpanzees often are absolutely identical in amino acid sequence (numbers of amino acid sites shown in parentheses): fibrinopeptide (30); cytochrome c (104); lysozyme (130); hemoglobin α (141); hemoglobin β (146); and hemoglobin δ (146). Some other closely similar proteins fully sequenced include hemoglobin γ (1 amino acid substitution among 146 sites), myoglobin (1 substitution, 153 sites), carbonic anhydrase (3 substitutions, 264 sites), serum albumin (6 substitutions, 580 sites), and transferrin (8 substitutions, 647 sites). Thus, altogether for these eleven proteins, a total of 2,468 of the 2,487 amino acid sites (99.2 percent) are identical between humans and chimpanzees.

22. For a recent summary, see the following and references therein: N. Takahata, "A Genetic Perspective on the Origin and History of Humans," *Annu. Rev. Ecol. Syst.* 26 (1995): 343–372.

23. S. Horai, K. Hayasaka, R. Kondo, K. Tsugane, and N. Takahata, "Recent African Origin of Modern Humans Revealed by Complete Sequences of Hominoid Mitochondrial DNAs," *Proc. Natl. Acad. Sci. USA* 92 (1995): 532–536.

24. As related species diverge from common ancestors, they tend to accumulate mutational differences at roughly regular rates that can be calibrated as follows. Living species with outstanding fossil or biogeographic records are assayed for molecular divergence under a particular laboratory method. Molecular evolutionary rates then are estimated by dividing the respective molecular divergence values by the times since common ancestry (as obtained from the independent biogeographic or fossil

evidence). Compilations of many such examples have shown sufficient rate regularities to have given rise to the concept of "molecular clocks" that apply to particular molecules, taxa, or assay procedures. These molecular clock calibrations can be employed to estimate provisional separation times for extant taxa when direct fossil evidence is poor.

25. C. G. Sibley and J. E. Ahlquist, "DNA Hybridization Evidence of Hominoid Phylogeny: Results from an Expanded Data Set," *J. Molec. Evol.* 26 (1987): 99–121; A. Caccone and J. R. Powell, "DNA Divergence among Hominoids," *Evolution* 43 (1989): 925–942.

26. As quoted in D. J. Futuyma, *Science on Trial* (New York: Pantheon Books, 1983).

27. This account is based on E. E. Max, "Plagiarized Errors and Molecular Genetics: Another Argument in the Evolution-Creation Controversy," *Creation/Evolution* XIX (1986): 34–46.

28. R. V. Collura and C.-B. Stewart, "Insertions and Duplications of mtDNA in the Nuclear Genomes of Old World Monkeys and Hominoids," *Nature* 378 (1995): 485–489.

29. Exact relationships among the various forms of *Australopithecus* and *Homo* known from fossils still are debated. Some hominid forms may have been in the direct line of descent leading to extant humans, whereas others may have been dead-end side branches of the evolutionary tree. However, even if the chain of ancestry of *Homo sapiens* was known precisely, room would remain for debate about where taxonomic boundaries should be drawn. In principle, any boundaries within a temporal continuum to some extent are arbitrary (every individual alive today had parents, who in turn had parents and so on back through time).

30. For some recent reviews and introductions to the literature, see the following: M. Nei and A. K. Roychoudhury, "Genetic Relationship and Evolution of Human Races," *Evol. Biol.* 14 (1982): 1–59; L. L. Cavalli-Sforza, P. Menozzi, and A. Piazza, *The History and Geography of Human Genes* (Princeton, N.J.: Princeton University Press, 1994).

31. This conclusion is supported further by a recent report on the ethnic distributions of more than a hundred genetic polymorphisms assayed by direct DNA-level (rather than protein-level) laboratory techniques: G. Barbujani, A. Magagni, E. Minch, and L. L. Cavalli-Sforza, "An Apportionment of Human DNA Diversity," *Proc. Natl. Acad. Sci. USA* 94 (1997): 4516–4519.

32. J. C. Avise, "Nature's Family Archives," *Natural History* 3 (1989): 24–27.

33. Family names first were used in China during the Han dynasty (about the time of Christ). However, the use of surnames to record family lines came much later to most of the world. Surnames were not customary in England until at least the fourteenth century. In Japan, only governing classes were allowed surnames until 1875, when cabinet decree mandated their adoption by the general populace.

34. A seminal study was by: W. M. Brown, "Polymorphism in Mitochondrial

DNA of Humans as Revealed by Restriction Endonuclease Analysis," *Proc. Natl. Acad. Sci. USA* 77 (1980): 3605–3609. Important recent extensions and reviews of this evidence include R. L. Cann et al., "Mitochondrial DNA and Human Evolution," *Nature* 325 (1987): 31–36; F. J. Ayala, "The Myth of Eve: Molecular Biology and Human Origins," *Science* 270 (1995): 1930–1936; N. Takahata, "A Genetic Perspective on the Origin and History of Humans," *Annu. Rev. Ecol. Syst.* 26 (1995): 343–372.

35. M. Nei and N. Takezaki, "The Root of the Phylogenetic Tree of Human Populations," *Molec. Biol. Evol.* 13 (1996): 170–177; S. A. Tishkoff et al., "Global Patterns of Linkage Disequilibrium at the CD4 Locus and Modern Human Origins," *Science* 271 (1996): 1380–1387.

36. See, for example, R. R. Hudson, "Gene Genealogies and the Coalescent Process," *Oxford Surveys Evol. Biol.* 7 (1990): 1–44.

37. One interesting ramification of this finding is that particular pieces of the genome in some humans truly are closer, genealogically speaking, to certain of those in chimpanzees than they are to homologous DNA sequences in some other human beings!

38. F. J. Ayala, "The Myth of Eve: Molecular Biology and Human Origins," *Science* 270 (1995): 1930–1936.

39. R. L. Dorit, H. Akashi, and W. Gilbert, "Absence of Polymorphism at the ZFY Locus on the Human Y Chromosome," *Science* 268 (1995): 1183–1185. Other recent studies of Y-chromosome genes are: M. F. Hammer, "A Recent Common Ancestry for Human Y Chromosomes," *Nature* 378 (1995): 376–378; M. F. Hammer et al., "The Geographic Distribution of Human Y Chromosome Variation," *Genetics* 145 (1997): 787–805; and L. S. Whitfield, J. E. Sulston, and P. N. Goodfellow, "Sequence Variation of the Human Y Chromosome," *Nature* 378 (1995): 379–380. These studies and reanalyses of the ZFY genetic data generally are consistent in suggesting that the father of all extant human Y chromosomes lived some 200,000 years ago, although the 95 percent statistical confidence limits surrounding such estimates are large (typically from 50,000 to 500,000 years).

40. J. C. Avise, "Mitochondrial DNA Polymorphism and a Connection between Genetics and Demography of Relevance to Conservation," *Conserv. Biol.* 9 (1995): 686–690. This paper points out that the number (as tallied by gender) of transmission pathways through a pedigree collectively available to most genes is $2^{(G+1)}$, where G is the number of generations. By contrast, the number of transmission pathways similarly tallied for mtDNA (or for the Y chromosome) is only 1. From this perspective, after even a few generations the total proportion of the hereditary history of a species "captured" by mtDNA (or the Y chromosome) is only a minuscule fraction ($1 / 2^{(G+1)}$) of the total.

41. Such studies are underway, and an interesting preliminary result is the formerly underappreciated complexity of the historical geographic patterns revealed by

different genes. See A. Gibbons, "Ideas on Human Origins Evolve at Anthropology Gathering," *Science* 276 (1997): 535–536.

42. Not everyone has agreed with this sentiment, noting instead some limitations and even potential dangers in the HGDP. For an introduction to the issues, see S. Lehrman, "Diversity Project: Cavalli-Sforza Answers His Critics," *Nature* 381 (1996): 14.

43. Actually, cells first were named and described by Robert Hooke in 1665 after microscopic examination (30X power) of bark from an oak tree. At about the same time, Antonie van Leeuwenhoek discovered not only blood cells and sperm cells in multicellular animals, but also entire cellular microbial worlds in microscopic examinations of droplets of pond water.

44. Use of the word "theory" here, though conventional, can be misleading if it is interpreted to imply merely an idle idea or possibility. A well-supported scientific theory is a conceptual scheme with broad explanatory and predictive power. The cell theory for life on Earth is every bit as well-founded empirically and conceptually as is the "theory" of gravity, the "theory" of a spherical (as opposed to a flat) earth, or the "theory" of evolution.

45. Some animal species, including a few vertebrates, reproduce by various asexual or quasi-sexual means that do not involve sperm-egg union. For example, a few lizards and fishes display parthenogenetic reproductive modes in which unreduced (diploid) eggs divide to give rise to new individuals without the benefit of functional genetic participation by sperm.

46. This pattern of radial cellular cleavage, in which the upper tier of four cells is aligned with the lower tier, is characteristic of all deuterostome animals, which in addition to chordates (creatures with backbones) include echinoderms (sea stars, brittle stars, sea cucumbers, sea urchins, and crinoids). By contrast, eight-celled embryos of protostomes (annelids, arthropods, mollusks, and many other invertebrates) display spiral cleavage in which the cells in the upper tier typically sit in the grooves between those of the lower echelon. These differences in early development provide one line of evidence for a phylogenetic separation distinguishing deuterostome animals from the protostomes.

47. This conclusion cannot be final because in some other species, exceptions exist to the generality that the genomes of all somatic cells in an individual are identical. In some insects, for example, amplification of rRNA genes occurs specifically in oocyte cells, and whole chromosomes or parts thereof can be eliminated from certain cells early in embryonic development. Also, red blood cells in humans contain no nuclei. Mutations are another source of genetic nonidentity for somatic cells.

48. M. D. Adams et al., "Initial Assessment of Human Gene Diversity and Expression Patterns Based upon 83 Million Nucleotides of cDNA Sequence," *Nature* 377S (1995): 3–174. The basic approach in this study was to isolate from various

tissues as many messenger RNA molecules as possible (these represent the products of actively expressed genes), reverse-transcribe these RNAs to their complementary ("c") DNAs, and directly sequence and thereby characterize the cDNAs.

49. In some organismal groups such as mollusks, blastomeres arise from asymmetrical divisions of a strongly polarized egg, and from the outset appear rigidly destined to form specific parts of the embryo.

50. Morphogens are chemical substances that vary in concentration along a gradient and provide information to a cell about that cell's location. For example, one recently discovered morphogen is retinoic acid, which in chick embryos displays concentration gradients that influence digit formation from limb buds.

51. C. Nüsslein-Volhard, "Gradients that Organize Embryo Development," *Sci. Amer.* 275 (1996): 54–61.

52. R. C. Duke, D. M. Ojcius, and J. D.-E. Young, "Cell Suicide in Health and Disease," *Sci. Amer.* 275 (1996): 80–87.

53. For example, during the Cambrian geologic period that began about 550 million years ago, a great profusion of invertebrate life with body forms even more diverse than those among living phyla today appeared in rapid order, (in no more than 10 million years). S. J. Gould, *Wonderful Life: The Burgess Shale and the Nature of History* (New York: Norton, 1989). Developmental alterations mediated by changes in regulatory genes almost certainly were involved in these evolutionary transformations. Among many treatments that elaborate the thesis of a developmental/regulatory connection, two early books were particularly influential: S. J. Gould, *Ontogeny and Phylogeny* (Cambridge, Mass.: Harvard University Press, 1977); and R. A. Raff and T. Kaufman, *Embryos, Genes, and Evolution* (New York: Macmillan, 1983).

54. The account here, including the automobile analogy, is taken from R. Tijan, "Molecular Machines that Control Genes," *Sci. Amer.* 272 (1995): 54–61.

55. Y. Muragaki, S. Mundlos, J. Upton, and B. R. Olsen, "Altered Growth and Branching Patterns in Synpolydactyly Caused by Mutations in HOXD13," *Science* 272 (1996): 548–551; D. P. Mortlock and J. W. Innis, "Mutation of HOXA13 in Hand-Foot-Genital Syndrome," *Nature Genetics* 15 (1997): 179–180. For a review of the evolutionary history of homeotic genes, see S. B. Carroll, "Homeotic Genes and the Evolution of Arthropods and Chordates," *Nature* 376 (1995): 479–485.

56. The current view can be termed epigenesis, the gradual appearance of new structures and functions during an organism's development. This contrasts to earlier "preformation" hypotheses in which all organismal features were thought to be present in miniaturized form in the egg.

57. Thus, one day on the cosmic calendar equals approximately 41 million years; one hour, 1.7 million years; one minute, 28,000 years; and one second, 500 years.

58. As quoted in D. J. Futuyma, *Science on Trial* (New York: Pantheon Books, 1983).

3. Genetic Maladies

1. A recent biography of Sir Archibald Garrod is A. G. Beam, *Archibald Garrod and the Individuality of Man* (New York: Oxford University Press, 1993).

2. Given the huge numbers of DNA polymorphisms already observed in human populations, the number of potential genotypes is astronomical. Suppose that a mere two hundred polymorphisms were present (more than an order of magnitude fewer than have been documented to date), each with the minimum possible two alleles. Under the logic of Mendelian heredity, the number of distinct genotypes that in principle could be generated by the shuffling action of recombination is 3^{200}. About 6,000,000,000 humans currently inhabit the Earth. Thus, only a minuscule fraction of the available human "genotypic space" is realized, and the probability of joint occupancy of any given genotypic slot is infinitesimally low.

3. Symptoms of the alkaptonuric condition also can be precipitated in nonalkaptonuric patients by the prolonged exposure to carbolic acid dressings for chronic cutaneous ulcers. This example of a phenocopy (a nongenetically produced phenotype resembling a genetically determined one) highlights how some metabolic disorders can have both genetic and environmental etiologies. In this case, the symptoms stem from an imbalance between the environmental challenge and the genetically-based capacity for appropriate response.

4. C. R. Scriver, A. L. Beaudet, W. S. Sly, and D. Valle, eds., *The Metabolic and Molecular Bases of Inherited Disease,* 7th ed. (New York: McGraw-Hill, 1995). V. A. McKusick, ed., *Mendelian Inheritance in Man,* 11th ed. (Baltimore, Md.: Johns Hopkins University Press, 1994); the online update can be accessed via the World Wide Web at http://www3.ncbi.nlm.nih.gov/omim/.

5. C. R. D. Brothers, "Huntington's Chorea in Victoria and Tasmania," *J. Neurol. Sci.* 1 (1964): 405–420.

6. N. S. Wexler et al., "Homozygotes for Huntington's Disease," *Nature* 326 (1987): 194–197.

7. J. F. Gusella et al., "A Polymorphic DNA Marker Genetically Linked to Huntington's Disease," *Nature* 306 (1983): 234–238.

8. Other disease genes with unstable numbers of short (trinucleotide) sequence repeat motifs include the fragile X syndrome (a form of mental retardation produced by an allele on the X chromosome), Kennedy syndrome (a motor neuron disease transmitted as a recessive allele also on the X), spinocerebellar ataxia type 1 (neurologic disorders with dominant allelic transmission on chromosome 6), dentatorubro-pallidoluysian atrophy (a neurodegenerative disorder due to a dominant allele on chromosome 12), and myotonic dystrophy (a progressive disorder of muscle weakness encoded by a dominant allele on chromosome 19). The unstable expansion of trinucleotide repeats at such "microsatellite genes" may be a common denominator for many dominantly inherited neurodegenerative disorders. See C. T. Ashley,

Jr. and S. T. Warren, "Trinucleotide Repeat Expansion and Human Disease," *Annu. Rev. Genetics* 29 (1995): 703–728; S. Karlin and C. Burge, "Trinucleotide Repeats and Long Homopeptides in Genes and Proteins Associated with Nervous System Disease and Development," *Proc. Natl. Acad. Sci. USA* 93 (1996): 1560–1565.

"Minisatellite loci" constitute another class of genetic elements with tandem-repeat motifs, but the repeated units are longer individually (typically fourteen to a hundred nucleotides each). Instabilities in minisatellite repeat numbers at some genes, such as the Ha-ras and insulin loci, contribute to the heritable risk of various cancers (carcinomas of the breast, the colon, the urinary bladder, and acute leukemia), and insulin-dependent diabetes, respectively. See T. G. Krontiris, "Minisatellites and Human Disease," *Science* 269 (1995): 1682–1683.

9. N. S. Wexler, "The Tiresias Complex: Huntington's Disease as a Paradigm of Testing for Late-Onset Disorders," *FASEB J.* 6 (1992): 2820–2825.

10. N. A. Campbell, *Biology,* 2nd ed. (Redwood City, Calif.: Benjamin-Cummings, 1990), p. 247.

11. The nucleotide substitution is a guanine for an adenine at position 578 in the gene encoding the sixth transmembrane helix of the receptor for leutenizing hormone.

12. The clinical presentation in this gender also can depend, however, on the degree of skewness in the pattern of X-chromosome inactivation. In female mammals, including humans, one of the two X chromosomes in each somatic cell is inactivated early in embryonic development. This phenomenon, known as the "Lyon effect" after Mary Lyon, who first described the process in mice in 1961, is related to the metabolic desirability of compensating for the gene dosage differences that otherwise would exist between females (with two copies of the X per cell) and males (with only one copy).

13. The pathways involved are oxidative phosphorylation and electron transport. These pathways result in the production of molecules of ATP (adenosine triphosphate) that act like batteries in storing cellular energy (in this case, in phosphate bonds).

14. Although mtDNA is transmitted almost exclusively from the mother, instances are known in other species where mtDNA molecules in offspring occasionally trace to their father's sperm, a phenomenon referred to as paternal leakage.

15. Two reviews are D. C. Wallace, "Mitochondrial Genetics: A Paradigm for Aging and Degenerative Diseases?" *Science* 256 (1992): 628–632; and D. C. Wallace, "Diseases of the Mitochondrial DNA," *Annu. Rev. Biochem.* 61 (1992): 1175–1212.

16. Free radical or oxygen radical molecules are destructive because they lack an electron, and this makes them prone to snatch electrons from other molecules (oxidation).

17. The distinction between simple and multifactorial genetic diseases is to some extent definitional. For example, cardiovascular diseases collectively arise from al-

terations in any of a multitude of genes that influence the morphogenesis and physiology of the circulatory system, often in combination with environmental stresses. Yet, particular cardiovascular disorders such as DiGeorge syndrome sometimes can be attributed to specifiable single-gene mutations that produce symptoms so characteristic as to warrant recognition as a distinct clinical syndrome.

18. W. K. Cavenee and R. L. White, "The Genetic Basis of Cancer," *Sci. Amer.* 272 (1995): 72–79.

19. For example, a gene known as p53 resides on the short arm of human chromosome 17. Known to be involved in more than fifty different cancers when mutated, this gene in its normal state produces a protein that functions in several biochemical pathways, including those involved in the repair of DNA damages and the suppression of tumors. Because of its medical importance, p53 won the "Molecule of the Year Award" from *Science* magazine (a sort of Academy Award in the scientific world).

20. Cellular genes that control normal proliferation of cells are referred to as proto-oncogenes. A proto-oncogene can be converted to an oncogene (carcinogenic form) by a somatic mutation, or by contact with a virus. Viruses that promote cancerous growth in humans do so either by introducing to cells altered forms of human genes picked up during their infective travels, or by activating host genes that otherwise are quiescent.

21. A recent issue of *Science* [272 (1996)] devoted to cardiovascular disease provides a useful introduction to this topic.

22. The last human in the world to contract smallpox was a Somalian man, Ali Maow Maalin. The year was 1977. It had taken nearly two centuries, following Edward Jenner's demonstration in 1796 of the efficacy of smallpox vaccination, for the world to be rid of this dreaded affliction. For a history of the victory over smallpox, see M. Pines, ed., *The Race Against Lethal Microbes,* (Chevy Chase, Md.: Howard Hughes Medical Institute, 1996).

23. In *Biology as Ideology: The Doctrine of DNA* (New York: Harper Perennial, 1992), Richard Lewontin suggests that the tubercle bacillus is necessary but not sufficient for tuberculosis. He notes that the disease was common in sweatshops and factories of the nineteenth century, but much rarer among country people and the upper classes. He concludes "we might be justified in claiming that *the* cause of tuberculosis is unregulated industrial capitalism, and if we did away with that system of social organization, we would not need to worry about the tubercle bacillus. When we look at the history of health and disease in modern Europe, that explanation makes at least as good sense as blaming the poor bacterium." Such reasoning, though unorthodox, nonetheless serves to emphasize that many disease *agents* such as the tubercle bacillus are only a component of a broader nexus of causality that can include environmental factors that facilitate the action of the proximate agent itself.

24. AIDS (acquired immune deficiency syndrome) is merely among the latest of

infectious pandemics with considerable demographic impact on human populations. The following describe human genetic variation in susceptibility to infection by the AIDS virus: M. Samson et al., "Resistance to HIV-1 Infection in Caucasian Individuals Bearing Mutant Alleles of the CCR-5 Chemokine Receptor Gene," *Nature* 382 (1996): 722–725; R. Liu et al., "Homozygous Defect in HIV-1 Coreceptor Accounts for Resistance of some Multiply-Exposed Individuals to HIV-1 Infection," *Cell* 86 (1996): 367–377.

25. D. J. Weatherall, *The New Genetics and Clinical Practice,* 2nd ed. (New York: Oxford University Press, 1985) as reported in E. H. McConkey, *Human Genetics, the Molecular Revolution* (Boston, Mass.: Jones & Bartlett, 1993).

26. At a balance between mutation (at rate m) to a deleterious allele, and purifying selection (as quantified by the relative selection intensity [s] against the defective genotypes), the equilibrium frequency of a detrimental allele in a population is given by $q = (m/s)^{1/2}$ if the allele is completely recessive, or by $q = m/hs$ if the allele is partially dominant (where h is the degree of dominance).

27. C. Ruwende et al., "Natural Selection of Hemi- and Heterozygotes for G6PD Deficiency in Africa by Resistance to Severe Malaria," *Nature* 376 (1995): 246–249.

28. Let s and t designate the relative selection intensities against A/A and S/S homozygotes, respectively. At equilibrium, the predicted frequency of the S allele is given by $q = s/(s + t)$, and that of the A allele by $t/(s + t)$. For example, if S/S homozygosity is lethal ($t = 1$), and A/A homozygosity diminishes fitness by 50 percent on average relative to the heterozygous condition ($s = 0.5$), then the expected equilibrium frequency of the S allele is $q = 0.33$. Thus, frequencies attainable for variant alleles under overdominant selection (selection in which heterozygotes have higher fitness than either homozygote) can be much higher than those typically associated with recurrent mutation alone.

4. Genetic Beneficence

1. J. D. Watson and F. H. C. Crick, "A Structure for Deoxyribose Nucleic Acid," *Nature* 25 (1953): 737–738.

2. L. M. Adleman, "Molecular Computation of Solutions to Combinational Problems," *Science* 266 (1994): 1021–1024; for further thought on this topic, see R. J. Lipton, "DNA Solution of Hard Computational Problems," *Science* 268 (1995): 542–545.

3. Some researchers also envision proteins as promising computational molecules, perhaps to be employed in conjunction with conventional semiconductors in a hybrid computer technology. See R. R. Birge, "Protein-Based Computers," *Sci. Amer.* 272 (1995): 90–95.

4. D. M. Hunt, K. S. Dulai, J. K. Bowmaker, and J. D. Mollon, "The Chem-

istry of John Dalton's Color Blindness," *Science* 267 (1995): 984–988. For a detailed description of X-linked genes encoding photosensitive pigments of the retina, see M. Neitz and J. Neitz, "Numbers and Ratios of Visual Pigment Genes for Normal Red-Green Color Vision," *Science* 267 (1995): 1013–1016.

5. In the 1940s, George Beadle and Edward Tatum conducted pioneering biochemical experiments on the bread mold *(Neurospora crassa)* that led to the "one-gene, one-enzyme" hypothesis. See "Genetic Control of Biochemical Reactions in *Neurospora*," *Proc. Natl. Acad. Sci. USA* 27 (1941): 499–506. These widely applicable discoveries showed that many genes code for particular stretches of amino acid sequence corresponding to a functional enzyme or major subunit thereof.

6. In the last few decades, with the advent of more direct procedures for molecular genetic assay (such as DNA and protein sequencing), the former requisite of genetic variation for gene detection has become somewhat relaxed. Particular genes and their protein products now can be identified provisionally by structure even when they fail to display assayable variation. Nonetheless, as evidenced by the descriptions of numerous genetic disorders in Chapter 3, the molecular search for a particular gene usually is prompted and certainly facilitated by the presence of genetically-based variation in the morphological or physiological conditions to which the gene contributes.

7. By alternative accounting criteria, many more genes than this already may be known. The Institute for Genome Research in Maryland pioneered an approach of "whole genome shotgun sequencing," whereby chromosomes are broken into tiny pieces by sound waves, the fragments sequenced, and the pieces in effect then put back together or realigned in computer searches for homologous overlaps. From this approach, partial sequences from more than 30,000 human genes recently were published in a 379-page developing atlas of the human genome [*Nature* 377S (1995): 1–379]. Of course, most of these partially sequenced genes will require further characterization before their full structures and possible functions are understood.

8. For discussions about how average gene size has been estimated, see Chapter 2 in E. H. McConkey, *Human Genetics* (Boston: Jones & Bartlett, 1993).

9. The presence of alpha repeats near centromeres (chromosomal sites of attachment for cellular fibers that direct the chromosomes to appropriate positions during cell division) suggests a role for these sequences in chromosomal alignment, but this speculation has not been verified.

10. C. Dib et al., "A Comprehensive Genetic Map of the Human Genome Based on 5,264 Microsatellites," *Nature* 380 (1996): 152–154.

11. During the long-term evolutionary process, nearly all genes ultimately must have arisen through duplications of preexisting genetic material as noted many years ago by S. Ohno, *Evolution by Gene Duplication* (New York: Springer-Verlag, 1970).

12. Following transcription of the entire gene (formation of messenger RNA from DNA), RNA processing enzymes typically remove the transcript sequences

corresponding to introns, and splice together exon sequences to produce a mature mRNA that subsequently is translated to protein. Alternative splicing can contribute to protein diversity during human development, and is one reason why different mutations in a particular gene sometimes have different clinical consequences. Although introns usually do not qualify as functional genes *per se,* mutations within them can be deleterious, as, for example, when they delete normal splice sites. The first introns were discovered in 1977, and soon were made known widely by an important commentary: W. Gilbert, "Why Genes in Pieces?" *Nature* 271 (1978): 501. Introns appear to be ubiquitous in the genomes of eukaryotic organisms. Much current debate centers on whether introns were present at the evolutionary outset in eukaryotic genes (the introns-early hypothesis), or whether they invaded eukaryotes much later (introns-late). See J. S. Mattick, "Introns: Evolution and Function," *Current Biol.* 4 (1994): 823–831.

13. The term genomic complexity has a specific meaning in molecular biology: the total length of *different* DNA sequences in the nucleus of a cell. Thus, a genome with a high fraction of highly repetitive DNA has lower sequence complexity than a comparably sized genome composed primarily of functional single-copy genes.

14. S. Brenner, G. Elgar, R. Sandford, A. Macrae, B. Venkatesh, and S. Aparicio, "Characterization of the Pufferfish *(Fugu)* Genome as a Compact Model Vertebrate Genome," *Nature* 366 (1993): 265–268.

15. Interestingly, the total gene number in humans has been of pragmatic concern to corporate backers of the entrepreneurial biotechnology firms that contemplate selling patent rights to genes and marketing gene products. See the discussion in *Science* 275 (1997): 769.

16. This is certainly a troublesome question, since, in many religions, human essences and aspirations (souls and heavens) lie within the metaphysical realm.

17. The closest known rRNA gene sequence affiliations of mammalian mtDNA lie within the α–subdivision of the purple photosynthetic bacteria.

18. Symbiosis describes the situation in which dissimilar organisms live together in close association. "Endo," derived from the Greek "endon" meaning within, has been prefixed to "symbiosis" in reference to the particularly intimate associations between microbes that attended the formation of the first eukaryotic cells.

19. Similar statements about endosymbiotic origins can be made for the chloroplast genomes of plants, which bear a strong resemblance to a different group of primitive bacteria. In general, prokaryotes are microorganisms that lack a membrane-bound nucleus containing chromosomes. They are to be distinguished from eukaryotes, which are all multicellular animals and plants as well as some unicellular organisms such as Protozoa.

20. Recent genetic evidence shows that mitochondrial → nuclear DNA tranfers continue today, although none of the recently transferred mitochondrial genes or gene pieces is thought to be functionally operative in the nucleus. See D.-X. Zhang

and G. M. Hewitt, "Nuclear Integrations: Challenges for Mitochondrial DNA Markers," *Trends Ecol. Evol.* 11 (1996): 247–251. One contributing explanation may be that following the original endosymbiotic event(s) hundreds of millions of years ago, evolutionary alterations in the genetic code apparently took place. This has been gauged by the slightly different mtDNA codes currently used by some deeply separated animal groups. For example, the code employed by vertebrate mtDNA differs slightly from that of the "universal" code of nuclear DNA (as well as that of various invertebrate groups). Thus, even if the DNA sequence for an entire human mitochondrial gene were transported to the nucleus now, any of its mRNA transcripts presumably would be recognized improperly by the nuclear protein-synthesizing machinery.

21. D. C. Wallace, "Mitochondrial Genes and Disease," *Hospital Practice* 21 (1986): 77–92.

22. Theodosius Dobzhansky, a well-known evolutionary biologist, is often quoted as saying: "Nothing in biology makes sense except in the light of evolution" [*Amer. Biol. Teacher* 35 (1973): 125–129].

23. The phrase "evolutionary tinkering" was used in an essay of the same name by F. Jacob in *The Possible and the Actual* (New York: Pantheon, 1982). In this essay are many additional examples of evolutionary puttering with organismal design.

24. Of course, not everyone agrees. For a recent book that sees the hand of a god in all the details of organismal biochemical operation, see M. J. Behe, *Darwin's Black Box: The Biochemical Challenge to Evolution* (New York: Simon & Schuster, 1996). For a critique of Behe's book more in line with my argument, see J. A. Coyne, "God in the Details," *Nature* 383 (1996): 227–228.

25. Such issues also arise in many other areas of scientific evidence against recent "special creation." If evolution had not taken place, why would a creator have gone to such effort to plant so many fossils, provide misleading evidence in radiocarbon dating, or dupe geneticists by leaving abundant evidence of historical legacy in the genomes of extant organisms? Unless most of physics, astronomy, geology, biology, and other sciences are entirely wrongheaded, any god that was a recent special creationist must also be a great prankster. Robert Frost said it thus: "Lord forgive all the little tricks I play on you, and I'll forgive the great big one you played on me."

26. One example of alternative splicing involves a regulatory muscle protein in mammals called troponin T. As a result of RNA processing events that bring together alternative exon domains from the processed RNA transcript, different forms of the protein appear at different stages in muscle development and differentiation.

27. In the 1940s, Richard Goldschmidt challenged conventional evolutionary wisdom by claiming that morphological evolution proceeds largely in discontinuous fashion, driven by developmental macromutations with dramatic phenotypic effects. Most evolutionists scoffed at Goldschmidt's "hopeful monsters," deeming them unlikely to find suitable niches in nature and survive the rigors of natural selection.

However, in the 1970s a resurgence of sorts brought ideas of saltational evolution back into the limelight. In particular, Allan Wilson and colleagues at the University of California at Berkeley, while not advocating Goldschmidt's extreme views, nonetheless emphasized the important role that altered patterns in gene expression could play in promoting the rapid evolution of morphological and behavioral features. For a review, see A. C. Wilson, "The Molecular Basis of Evolution," *Sci. Amer.* 253 (1985): 164–173.

28. Over the broad sweep of evolution from relatively simple microbes to complex multicellular eukaryotes, gene duplication processes clearly have been of paramount importance in permitting expansion of both the size and the functional diversity of organismal genomes. See also note 11.

29. N. A. Campbell, *Biology,* 2nd ed. (New York: Benjamin/Cummings, 1990).

30. S. A. Kauffman, *At Home in the Universe* (New York: Oxford University Press, 1995).

31. C. Wills, *The Wisdom of the Genes* (New York: Basic Books, 1989).

32. For an extended introduction to this line of inquiry, see A. A. Hoffmann and P. A. Parsons, *Evolutionary Genetics and Environmental Stress* (New York: Oxford University Press, 1991).

33. *Philosophie Zoologique, ou Exposition des Considérations Relatives a l'Histoire Naturelle des Animaux* (Paris, 1809).

34. In his 1868 book entitled *The Variation of Animals and Plants under Domestication,* Charles Darwin advanced a "provisional hypothesis of pangenesis" which proposed that each part of the body produces minute hereditary particles, or gemmules, some of which are transferred through the bloodstream and collected in eggs and sperm for transmission to progeny. For example, liver gemmules were produced by the liver and heart gemmules by the heart. This model, soon shown experimentally to be incorrect by the German biologist August Weismann, had been prompted by valid criticisms that Darwin had no explicit theory of heredity underlying his evolutionary ideas. Ironically, in retrospect the gemmule hypothesis would appear to be more compatible with Lamarckian inheritance than with the Mendelian modes of inheritance that later were to prove pivotal for Darwinian evolution.

35. Until recently, a "central dogma" of genetics held that the sole direction of information flow in cells was: DNA $\rightarrow$ RNA $\rightarrow$ protein. With the discovery of enzymes that can catalyze the formation of DNA from RNA, the diagram must be modified somewhat to DNA $\leftrightarrow$ RNA $\rightarrow$ protein. Nevertheless, there is as yet no evidence for a coding feedback to the hereditary material itself from proteins (much less from composite body parts such as giraffe necks, or eyes). At the present time, genetic variation arising from germline mutations (broadly construed to include point mutations, recombination, insertions of transposable elements and viruses, and other such sources of genetic alteration) constitutes the only known hereditary fodder for natural selection.

36. Adaptation is used here in an unusually broad sense to include "acclimation," "acclimatization," changes in regulatory or ontogenetic profiles, and other related processes that represent evolved capacities for organismal solutions to environmental problems encountered during individual lifetimes. The genes underlying such adaptive responses are not themselves heritably altered during the response, but any inherent variations in genes that influence the nature of the adaptational response can be selected and heritably transmitted to offspring.

37. The molecular and cellular biology of the immune response is vastly more complicated than described here and involves many other genes and cell types. For a comprehensive introduction to the human immune response, see B. Lewin, *Genes VI* (New York: Oxford University Press, 1997).

38. Charles Darwin's book, *The Variation of Animals and Plants under Domestication,* describes many biotic responses to artificial selection known in the mid-1880s.

39. For introductions to peppers, see W. H. Greenlead, "Pepper Breeding," in M. J. Bassett, ed., *Breeding Vegetable Crops* (Westport, Conn.: Avi Publishing Co., 1986), pp. 67–134; A. Naj, *Peppers: A Story of Hot Pursuits* (New York: Alfred A. Knopf, 1992).

40. The association of dogs with man began some 10,000 years ago, but many spectacular and specialized of the more than 400 distinct breeds in the world today have been generated selectively in just the past few decades and centuries.

41. See L. H. Rieseberg and S. M. Swensen, "Conservation Genetics of Endangered Island Plants," in J. C. Avise and J. L. Hamrick, eds., *Conservation Genetics: Case Histories from Nature* (New York: Chapman & Hall, 1996), pp. 305–334.

42. See S. J. O'Brien et al., "Conservation Genetics of the Felidae," in Avise and Hamrick, *Conservation Genetics*, pp. 50–74.

43. In one sense, the evolutionary stasis of horseshoe crabs is illusory only. At the level of proteins and DNA, horseshoe crabs display normal levels of genetic variation and rates of molecular evolutionary change. See R. K. Selander, S. Y. Yang, R. C. Lewontin, and W. E. Johnson, "Genetic Variation in the Horseshoe Crab *(Limulus polyphemus),* a Phylogenetic 'Relic,'" *Evolution* 24 (1970): 402–414; J. C. Avise, W. S. Nelson, and H. Sugita, "A Speciational History of 'Living Fossils': Molecular Evolutionary Patterns in Horseshoe Crabs," *Evolution* 48 (1994): 1986–2001.

5. Strategies of the Genes

1. See *Theogony* and *Works and Days,* dating from the eighth century B.C.E., by the Greek poet Hesiod.

2. Few organisms are strictly clonal for long periods of time, however. Most bacterial strains, for example, occasionally undergo genetic exchange via: (a) conjugative transfer of chromosomal DNA; (b) the swap of snippets of DNA known as plasmids; (c) transformation, whereby DNA is taken up from the environment; or

(d) transduction, whereby DNA transfer is viral mediated. Most species of plants and invertebrate animals with regular asexuality also display sexual modes of reproduction at particular stages of the life cycle or under certain environmental conditions. Thus, the evolutionary destinies of separate genes within their genomes are decoupled partially.

3. Species of asexual vertebrates occupy outermost twigs on the evolutionary tree of life rather than major branches. Thus, they have evolved from sexual ancestors and tend to be evolutionarily short-lived. All known asexual vertebrate species arose through hybridization events between related sexual forms. See R. M. Dawley and J. P. Bogart, eds., *Evolution and Ecology of Unisexual Vertebrates* (Albany: State University of New York, 1989). Thus, for reasons of historical legacy, the genomes of asexual vertebrates may retain many of the properties (including some negative interactions among genes) displayed by sexual species.

4. One might suppose, for example, that all genes involved in the TCA metabolic pathway ideally should be inherited as a nonrecombining block, such that once proper functional interactions among the gene products had evolved, these advantages could be frozen in place for future generations. However, most functionally related loci in higher animals carry separate genomic addresses, and, accordingly, are sorted and transmitted independently under sexual reproduction. On the other hand, some tightly linked chromosomal "supergenes" are known wherein chromosomally adjacent and functionally interactive genes usually are transmitted together. One class of examples involves genetic sequences that regulate transcription, which typically are situated immediately adjacent to a relevant coding region, and hence (appropriately enough) tend to be cotransmitted with the alleles whose expression they govern.

5. H. J. Muller, "The Relation of Recombination to Mutational Advance," *Mutation Res.* 1 (1964): 2–9. The important idea raised in this paper was later coined "Muller's ratchet" by J. Felsenstein, "The Evolutionary Advantage of Recombination," *Genetics* 78 (1974): 737–756. Muller's ratchet ignores the possibility of rare back mutations to the nondeleterious state, or of counterbalancing beneficial mutations at other genes influencing fitness.

6. J. F. Crow, "The Importance of Recombination," in R. E. Michod and B. R. Levin, eds., *The Evolution of Sex* (Sunderland, Mass.: Sinauer, 1988), pp. 56–73. Calculations in this paper suggest that Muller's ratchet limits the size of genomes that can be transmitted by strictly asexual modes.

7. This term was coined by L. Sandler and E. Novitski, "Meiotic Drive as an Evolutionary Force," *Amer. Natur.* 91 (1957): 105–110.

8. All meiotic drive systems analyzed thus far involve the participation of at least two loci in a gene complex: a distortion locus, and a responder locus that the distorter targets. The mechanistic complexity of meiotic drive systems suggests for this reason alone that they might be relatively rare.

9. This phrase may have originated with E. G. Leigh, *Adaptation and Diversity* (San Francisco: Freeman, Cooper & Co., 1971).

10. Every generation of individuals, regardless of the population sex ratio, contains an equal number of autosomal alleles inherited from male and female parents. This fundamental truth led to the realization, first formalized in R. A. Fisher, *The Genetical Theory of Natural Selection* (Cambridge: Oxford University Press, 1930), that a form of frequency-dependent selection operates in such a way as to favor parental investment strategies that culminate in approximately equal sex ratios in populations. If males are infrequent in a population, families producing disproportionate numbers of sons (relative to their costs of production) will on average leave more grandchildren than do families that produce excess daughters, and, thus, any genes for male-producing tendencies spread. Conversely, if females are infrequent in a population, families producing disproportionate numbers of daughters (relative to their costs of production) will on average leave more grandchildren than do families that produce excess sons, and, thus, any genes for female-producing tendencies spread. In other words, autosomal genes tend to maximize their mean fitness by producing the minority sex, such that at equilibrium a 1:1 sex ratio in the population typically is approached. Over the long course of evolution, this selectively-favored equilibrium (from the perspective of autosomal genes) appears to have been genetically codified in many species of higher animals by the evolution of sex-determining chromosomes and a mechanistic meiotic process that assures a more or less equal collective production of males and females.

11. The prospect of capitalizing upon Y chromosome drive to control crop pests has not escaped the attention of agriculturists. If a forceful Y-driving chromosome could be engineered genetically and introduced into a pest population, it would perpetuate itself at the expense of the normal Y, thereby reducing the number of females and perhaps causing extinction of the blight. For a specific example and a broader discussion of Y-drive, see the following: T. W. Lyttle, "Experimental Population Genetics of Meiotic Drive Systems I. Pseudo-Y Chromosomal Drive as a Means of Eliminating Cage Populations of *Drosophila melanogaster*," *Genetics* 86 (1977): 413–445; and T. W. Lyttle, "Segregation Distorters," *Annu. Rev. Genet.* 25 (1991): 511–557.

12. W. D. Hamilton, "Extraordinary Sex Ratios," *Science* 156 (1967): 477–488.

13. An X-linked driver allele that forces excess production of X-bearing sperm (and hence females) would not profit from this genic selection pressure when present in females, where it spends two-thirds of its evolutionary lifetime. The excess evolutionary time spent in females comes about because every organismal generation has an equal number of male and female parents (regardless of the sex ratio in the population), and each female has two copies of the X compared to the one copy in males.

14. See C.-I. Wu and M. F. Hammer, "Molecular Evolution of Ultraselfish

Genes of Meiotic Drive Systems," in R. K. Selander, A. G. Clark, and T. S. Whittam, eds., *Evolution at the Molecular Level* (Sunderland, Mass.: Sinauer, 1991), pp. 177–203.

15. This is not invariably true. In coniferous plants, chloroplast DNA is inherited paternally.

16. Population genetic models can describe the quantitative effects. At equilibrium, the expected frequency of a recurring mtDNA mutation is approximately $q = m/s_f$, where m is the mitochondrial mutation rate in the female germ line and $1 - s_f$ is the fitness of a female carrying the mutation relative to a normal female with a fitness of one [S. A. Frank and L. D. Hurst, "Mitochondria and Male Disease," *Nature* 383 (1996): 224]. Note that male fitness has no influence on the expectation. Thus, for example, a recurring mtDNA mutation with $m = 10^{-4}$ and $s_f = 0.01$ has an expected equilibrium frequency in the population of $q = 0.01$, regardless of whether the mutation improves the fitness of males, or causes them to die prematurely or be nearly sterile.

17. For summaries of this literature, see L. D. Hurst, "The Incidences, Mechanisms and Evolution of Cytoplasmic Sex Ratio Distorters in Animals," *Biol. Rev.* 68 (1993): 121–193; W. G. Eberhard, "Evolutionary Consequences of Intracellular Organelle Competition," *Quart. Rev. Biol.* 55 (1980): 231–249.

18. For further thought on this topic, see I. M. Hastings, "Population Genetic Aspects of Deleterious Cytoplasmic Genomes and their Effect on the Evolution of Sexual Reproduction," *Genet. Res. Camb.* 59 (1992): 215–225; and J. Maynard Smith and E. Szathmáry, *The Major Transitions in Evolution* (New York: W.H. Freeman & Co., 1995).

19. Mobile elements are of several types, and can be classified by various and sometimes overlapping criteria. One criterion distinguishes mobile elements that transpose directly as DNA (class II elements) from those that do so by reverse transcription of an RNA intermediate (class I). Another important criterion distinguishes replicative (proliferative) elements from nonreplicative ones that move about without increase in copy number. Another classification scheme focuses on comparative molecular structures, transposition mechanisms, and inferred phylogenies, and distinguishes "cut and paste" transposons, retrotransposons, LINEs (long interspersed elements), and mobile introns. For an elaboration of this latter scheme, as well as additional discussion about the evolutionary strategies of mobile DNA, see C. Zeyl and G. Bell, "Symbiotic DNA in Eukaryotic Genomes," *Trends Ecol. Evol.* 11 (1996): 10–15. For general background, see D. E. Berg and M. M. Howe, *Mobile Elements* (Washington, D.C.: American Society for Microbiology, 1989).

20. Early, influential papers advancing this view are W. F. Doolittle and C. Sapienza, "Selfish Genes, the Phenotype Paradigm and Genome Evolution," *Nature* 284 (1980): 601–603; and L. E. Orgel and F. H. C. Crick, "Selfish DNA: The Ultimate Parasite," *Nature* 284 (1980): 604–607.

21. M. E. Lambert, J. F. McDonald and I. B. Weinstein, eds., *Eukaryotic Transposable Elements as Mutagenic Agents* (New York: Cold Spring Harbor Laboratory, 1988). For an example of experimental studies that have documented deleterious consequences to the organism of the mutations induced by transposable element movements, see R. C. Woodruff, "Transposable DNA Elements and Life History Traits. 1. Transposition of P DNA Elements in Somatic Cells Reduces the Lifespan of *Drosophila melanogaster,*" *Genetica* 86 (1992): 143–154.

22. D. A. Hickey, "Selfish DNA, a Sexually-Transmitted Nuclear Parasite," *Genetics* 101 (1982): 519–531.

23. Y. Xiong and T. H. Eichbush, "Origin and Evolution of Retroelements Based upon Their Reverse Transcriptase Sequences," *EMBO J.* 9 (1990): 3353–3362.

24. Yet another line of educated speculation is that transposable elements and gene introns have intimate evolutionary relationships [review in M. D. Purugganan, "Transposable Elements as Introns: Evolutionary Connections," *Trends Ecol. Evol.* 8 (1993): 239–243]. One idea with considerable support is that introns originated as a result of mobile element insertion into coding sequences, and that the excision of intron sequences during mRNA processing represents one mechanism by which the deleterious effects of element insertions into genes are minimized. If mobile elements and most introns truly are related, then to the extent that exon shuffling and alternative splicing have played important roles in the evolutionary process, so too in this additional sense have mobile elements.

25. A good example is provided by jumping genes in fruit flies that cause hybrid dysgenesis, a syndrome of lethality and sterility in the progeny of particular crosses. Mobile P-elements responsible for the dysgenic effects transpose only in gonadal tissues.

26. See Berg and Howe, *Mobile Elements.*

27. This and the following statements apply only to mobile elements that are inherited strictly vertically (from parents to progeny). For mobile elements that are capable of horizontal transmission between hosts (such as some retroviruses), replicative proliferation can be selectively advantageous within populations of asexual as well as sexual hosts.

28. D. A. Hickey, "Evolutionary Dynamics of Transposable Elements in Prokaryotes and Eukaryotes," in J. F. McDonald, ed., *Transposable Elements and Evolution* (Dordrecht, The Netherlands: Kluwer, 1993), pp. 142–147.

29. Many of the LINE-1 elements are truncated from their full-length size of about 6.5 kilobases, and presumably are functionally immobile at present. This suggests that they are "fossilized" relicts from longer sequences formerly movable. On the other hand, *Alu* elements continue to transpose replicatively in contemporary human populations via the reverse transcription of RNA molecules stemming from at least three different source genes.

30. S. Langlois, S. Deeb, J. D. Brunzell, J. J. Kastelein, and M. R. Hayden, "A Major Insertion Accounts for a Significant Proportion of Mutations Underlying Human Lipoprotein Lipase Deficiency," *Proc. Natl. Acad. Sci. USA* 86 (1989): 948–952; B. A. Dombrowski, S. L. Mathias, E. Nanthakumar, A. F. Scott, and H. H. Zazazian, Jr., "Isolation of an Active Human Transposable Element," *Science* 254 (1991): 1805–1808; K. Muratani, T. Hada, Y. Yamamoto, T. Kaneko, Y. Shigeto, T. Ohue, J. Furuyama, and K. Higashino, "Inactivation of the Cholinesterase Gene by *Alu* Insertion: Possible Mechanism for Human Gene Transposition," *Proc. Natl. Acad. Sci. USA* 88 (1991): 11315–11319; M. R. Wallace, L. B. Anderson, A. M. Saulino, P. E. Gregory, T. W. Glover, and F. S. Collins, "A *de novo Alu* Insertion in Neurofibromatosis Type 1," *Nature* 353 (1991): 864–866.

31. The occasion was the annual meeting of the American Association for the Advancement of Science, held in Atlanta, Georgia in 1995, and the speaker was Dr. John McDonald from the University of Georgia. Two reviews by McDonald on the potential evolutionary significance of mobile elements are "Evolution and Consequences of Transposable Elements," *Current Opinion Genetics Develop.* 3 (1993): 855–864; and "Transposable Elements: Possible Catalysts of Organismic Evolution," *Trends Ecol. Evol.* 10 (1995): 123–126.

32. If so, such an outcome could have an interesting mix of Lamarckian aspects and those of traditional gene regulation. The phenomenon would qualify as Lamarckian to the extent that the "mutations" caused by mobile element insertions were adaptive and heritable (passed through the germ line), but the phenomenon could also be seen as non-Lamarckian if the "mutations" were confined to somatic cells and interpreted merely as regulatory responses of the genome to environmental signals.

33. R. Waterfield, *Plato Symposium, A New Translation* (New York: Oxford University Press, 1994).

34. One mechanistic reason for this in humans and other mammals is genomic imprinting, a phenomenon in which the genetic material from one sex is apparently labeled, during gametogenesis, as being different from its counterpart supplied by the opposite sex. Gender-specific methylation of DNA (whereby methyl groups are attached to carbon atoms primarily in cytosine residues) is one identified means by which genomic imprinting takes place. The net effect is that proper embryonic development requires the complementary interaction between a maternal and a paternal set of chromosomes. This is also one proximate reason why parthenogenetic development of an individual from two fused egg (or sperm) nuclei is unknown in mammals.

35. For example, female spiders sometimes eat their mates. Cannibalism aside, such dangers hardly are confined to nonhuman species. In the disease arena, for example, humans always have been plagued by a great variety of sexually transmitted or venereal diseases, of which AIDS is merely the latest to achieve notoriety.

36. D. A. Hickey and M. R. Rose, "The Role of Gene Transfer in the Evolution of Eukaryotic Sex," in R. E. Michod and B. R. Levin, eds., *The Evolution of Sex* (Sunderland, Mass.: Sinauer, 1988), pp. 161–175.

37. See chapters by Bernstein et al., Holliday, Levin, and Shields in Michod and Levin, *The Evolution of Sex*.

38. G. C. Williams, *Sex and Evolution* (Princeton, N.J.: Princeton University Press, 1975). See also chapters by Bell, Crow, Ghiselin, Maynard Smith, Seger and Hamilton, and others in Michod and Levin, *The Evolution of Sex*.

39. See, for example, J. Seger and W. D. Hamilton, "Parasites and Sex," in Michod and Levin, *The Evolution of Sex,* pp. 176–193.

40. C. Bernstein and H. Bernstein, *Aging, Sex,. and DNA Repair* (New York: Academic Press, 1991). The Bernsteins distinguish DNA damage from DNA mutation. The former converts normal double-stranded DNA to something else by such processes as strand breakage, crosslinking, or dimerization, whereas the latter merely alters one double-stranded DNA to another by such processes as nucleotide substitution, deletion, or inversion. For a popular account of the relationship between DNA damage repair and sexual reproduction, see R. E. Michod, *Eros and Evolution* (Reading, Mass: Addison-Wesley, 1995).

41. I. Mellon, D. K. Rajpal, M. Koi, C. R. Boland, and G. N. Champe, "Transcription-Coupled Repair Deficiency and Mutations in Human Mismatch Repair Genes," *Science* 272 (1996): 557–560.

42. Not all evolutionary geneticists agree. For an opposing view, see J. Maynard Smith, "The Evolution of Recombination," in Michod and Levin, *The Evolution of Sex,* pp. 106–125.

43. Sir Peter Medawar, a Nobel Prize–winning immunologist, used the test tube analogy in his development of aging concepts in a 1952 essay: *An Unsolved Problem of Biology* (London: H.K. Lewis). In this essay, Medawar clarified the evolutionary ramifications of the declining force of natural selection through successive age cohorts of a population, a realization that has become a cornerstone of evolutionary theories on aging and death.

44. Similar statements can be made for males, except for a noticeable hump in mortality rate curves for this gender between the ages of twelve and twenty-three. Steven Austad (*Why We Age: What Science is Discovering about the Body's Journey through Life* (New York: J. Wiley & Sons, 1997) refers to this as the time of testosterone dementia, when for behavioral rather than health reasons maturing boys are more than three times likelier to die than girls. The testosterone dementia that no doubt fostered enthusiastic warriors and hunters in days past today may predispose adolescent boys toward reckless behaviors that make young males poorer insurance risks than young females.

45. Notwithstanding biblical accounts of a 969-year-old Methuselah! Popular accounts still appear of humans living far beyond 130 years, but all remain unsub-

stantiated, and carefully researched claims invariably have proved to be mistakes or hoaxes. The 1995 Guinness Book of World Records lists 120 years as the oldest age confirmed for any human.

46. From the 1960s to the 1980s, the ideas originally formulated by Medawar were made more explicit and formal in a series of mathematical treatments by, among others, William Hamilton in Great Britain and Brian Charlesworth in the United States. The history of scientific thought on aging, as well as a summary of the theoretical treatments, can be found in M. R. Rose, *Evolutionary Biology of Aging* (New York: Oxford University Press, 1991).

47. This phrase, although conventional, was ill-chosen because it can be mistaken to imply the accumulation of deleterious somatic mutations during the lifetime of an individual (although these mutations also may contribute to the aging process). The traditional mutation-accumulation hypothesis invokes the evolutionary accumulation within populations of alleles with late-onset deleterious consequences to individuals. Thus, a more suitable moniker for this evolutionary scenario might be "age specificity of gene action."

48. Pleiotropy is a general term that refers to the propensity of individual genes to influence a number of distinct phenotypic features. In the current context, antagonistic pleiotropy implies that the same genes that provide fitness benefits to younger individuals may come at the expense of these same individuals when they grow older.

49. Medawar, *An Unsolved Problem of Biology.*

50. G. C. Williams, "Pleiotropy, Natural Selection, and the Evolution of Senescence," *Evolution* 11 (1957): 398–411.

51. In a sense, the DNA repair hypothesis is a subset of a broader model known as the disposable soma theory. See T. B. L. Kirkwood, "Evolution of Ageing," *Nature* 270 (1977): 301–304; T. B. L. Kirkwood and R. Holliday, "The Evolution of Ageing and Longevity," *Proc. Royal Soc. Lond B* 205 (1979): 531–546. Under the disposable soma theory, a tradeoff is envisioned for each organism between metabolic investment in the development and maintenance of the soma, and investment in the production of germ cells and other components of the reproductive system. A resolution favored by natural selection would optimize the survival of germ cells, implying that at some point in life the continued investment in soma becomes counterproductive. Thus, aging results from a curtailment of cellular mechanisms (including DNA repair) that otherwise might maintain a soma indefinitely.

52. The etiology of many cancers may provide a good example. Increasing evidence suggests that cancers often arise when normal controls on cellular proliferation are lost, and that such loss typically requires successive mutational hits at several genes. Some of these hit sites may have been inherited from one or another parent, thus predisposing an individual to cancer at some stage in life, but other mutational hits clearly arise during the lifetime of the individual. A single cell with

the appropriate (or, more correctly, inappropriate) suite of inherited and acquired mutations necessary for cancerous growth sometimes can spell quick doom for the individual organism.

53. The ensuing scenarios in the text, which build upon the Bernsteins' hypotheses concerning relationships between sexual reproduction and aging, are taken from J. C. Avise, "The Evolutionary Biology of Aging, Sexual Reproduction, and DNA Repair," *Evolution* 47 (1993): 1293–1301.

54. It comes as no surprise that baptisms, marriages, and funerals typically are conducted in houses of worship, with elaborate social ceremonies sanctioned to commemorate these rites of passage.

6. Genetic Sovereignty

1. Such possibilities certainly were not lost on Charles Darwin, who in 1872 published *The Expression of the Emotions in Man and Animals* (reprinted by the University of Chicago Press, 1965), a book that attempted to extend principles developed in *On the Origin of Species* to the fields of human ethology and psychology.

2. See R. C. Lewontin, S. Rose, and L. J. Kamin, *Not in Our Genes* (New York: Pantheon Books, 1984).

3. L. S. Hearnshaw, *Cyril Burt: Psychologist* (London: Hodder and Stoughton, 1979). The Burt episode and its place within the broader debate over genetics and human IQ also are reviewed in Lewontin, Rose, and Kamin, *Not in Our Genes,* and in S. J. Gould, *The Mismeasure of Man* (New York: Norton, 1981).

4. Between 1907 and 1917, sixteen American states passed sterilization laws directed against the mentally retarded. Constitutionality of these laws was upheld in a Supreme Court ruling in which Justice Oliver Wendell Holmes declared that "three generations of imbeciles are enough."

5. These were psychologist Robert Joynson and sociobiologist Ronald Fletcher, and their side of the story is recounted in another recent book on human intelligence and IQ: R. J. Herrnstein and C. Murray, *The Bell Curve* (New York: Simon & Schuster, 1994).

6. Such thoughts have been agonizing to me in composing this book. The subculture in which I have been trained accepts science as a means to understanding life. Prolonged exposure instead to Hinduism, Freudism, Marxism, or astrologism, would have given me quite different views. For a philosophical critique of the "legend" of science as an unerring approach to truth, see P. Kitcher, *The Advancement of Science* (New York: Oxford University Press, 1993).

7. In *The Mismeasure of Man*, S. J. Gould details the history of IQ research in the context of hereditarian theories.

8. Not all would agree with this statement. A fundamental empirical point emphasized in Herrnstein and Murray's *The Bell Curve* is that intelligence is real,

objective, and repeatably measurable by any of a battery of cognitive tests that all tend to converge on the same assessment and that also correlate well with common-sense perceptions of mental ability.

9. T. J. Bouchard, D. T. Lykken, M. McGue, N. L. Segal, and A. Tellegen, "Sources of Human Psychological Differences: The Minnesota Study of Twins Reared Apart," *Science* 250 (1990): 223–228.

10. The "broad-sense" heritability of a trait is defined as the proportion of the trait's total phenotypic variance within a population that is accounted for by genetic differences:

$$H = \text{genetic variance} / (\text{genetic variance} + \text{environmental variance}).$$

Because genetic and environmental variances are population-specific rather than universal properties, heritabilities for a trait can differ considerably from one setting to another as a result of differences in genetic variance, environmental variance, or both. Another caveat about particular heritability estimates reported in the literature is that they can vary according to the genetic model to which empirical data are fit. Some models, for example, assume additive allelic contributions by an unspecified number of genes (heritability in the "narrow sense"), whereas others may include nonadditive effects between alleles of a locus (dominance) or between alleles of different genes (epistasis). Furthermore, when extensive epistasis (gene interaction) is involved, twin studies may uncover evidence for strong genetic impact on variation in a trait, whereas parent-offspring or other sibship studies might suggest relatively low heritabilities for the same trait. Such discrepancies emphasize how genetic influences sometimes can be both substantial *and* nonfamilial.

11. D. Lykken and A. Tellegen, "Happiness is a Stochastic Phenomenon," *Psychological Sci.* 7 (1996): 186–189. In contrast to the large genetic influence on happiness or well-being, the authors report that none of the following could account for more than about 3 percent of the interperson variance in general sense of contentment: socioeconomic status, educational attainment, family income, marital status, or degree of religious commitment. Although each person's happiness fluctuates in response to life's contingencies, these transitory oscillations appear to center around a stable temperamental "set point" characteristic of each individual. See also D. G. Myers and E. Diener, "Who Is Happy?" *Psychological Sci.* 6 (1995): 10–19.

12. A. Tellegen et al., "Personality Similarity in Twins Reared Apart and Together," *J. Personality and Social Psychol.* 54 (1988): 1031–1039.

13. J. C. Loehlin and R. C. Nichols, *Heredity, Environment, and Personality: A Study of 850 Sets of Twins* (Austin, Tx.: University of Texas Press, 1976); J. P. Rushton, D. W. Fulker, M. C. Neale, D. K. B. Nias, and H. J. Eysenck, "Altruism and Aggression: The Heritability of Individual Differences," *J. Personality and Social Psychol.* 6 (1986): 1192–1198; N. G. Martin, L. J. Eaves, A. C. Heath, R. Jardine, L. M. Feingold, and H. J. Eysenck, "Transmission of Social Attitudes," *Proc. Natl.*

Acad. Sci. USA 83 (1986): 4364–4368; L. J. Eaves, H. J. Eyysenck, and N. G. Martin, *Genes, Culture and Personality: An Empirical Approach* (New York: Academic Press, 1989). A particularly intriguing twin study concluded that genetic factors account for about 60 percent of the variance in cognitive ability among the elderly [G. E. McClearn, B. Johansson, S. Berg, N. L. Pederson, F. Ahern, S. A. Petrill, and R. Plomin, "Substantial Genetic Influence on Cognitive Abilities in Twins 80 or More Years Old," *Science* 276 (1997): 1560–1563].

14. A quote attributed to Douglas Copeland in a recent *Rolling Stone* magazine.

15. Typical examples of numerous papers on this topic include V. S. Johnston and M. Franklin, "Is Beauty in the Eye of the Beholder?" *Ethology and Sociobiol.* 14 (1993): 183–199; S. W. Gangestad, R. Thornhill, and R. A. Yeo, "Facial Attractiveness, Developmental Stability, and Fluctuating Asymmetry," *Ethology and Sociobiol.* 15 (1994): 73–85. Several reviews also have appeared in scientific and popular outlets: J. Horgan, "The New Social Darwinists," *Sci. Amer.* 273 (1995): 174–181; G. Cowley, "The Biology of Beauty," *Newsweek,* 3 June 1996.

16. N. G. B. Jones and E. da Costa, "A Suggested Adaptive Value of Toddler Night Waking: Delaying the Birth of the Next Sibling," *Ethology and Sociobiol.* 8 (1987): 135–142; J. Shepher and J. Reisman, "Pornography: A Sociobiological Attempt at Understanding," *Ethology and Sociobiol.* 6 (1985): 103–114; J. W. Burgess, "Do Humans Show a 'Species-Typical' Group Size?" *Ethology and Sociobiol.* 5 (1984): 51–57; J. W. Burgess, "The Social Biology of Human Populations: Spontaneous Group Formation Conforms to Evolutionary Predictions of Adaptive Aggregation Patterns," *Ethology and Sociobiol.* 10 (1989): 343–359.

17. M. W. Weiderman and E. R. Allgeier, "Gender Differences in Sexual Jealousy: Adaptationist or Social Learning Explanation?" *Ethology and Sociobiol.* 14 (1993): 115–140; L. Paul and L. R. Hirsch, "Human Male Mating Strategies: 2. Moral Codes of 'Quality and Quantity' Strategists," *Ethology and Sociobiol.* 17 (1996): 71–86; K. B. Kerber, "The Marital Balance of Power and quid pro quo: An Evolutionary Perspective," *Ethology and Sociobiol.* 15 (1994): 283–297; D. Thiessen, R. K. Young, and R. Burroughs, "Lonely Hearts Advertisements Reflect Sexually Dimorphic Mating Strategies," *Ethology and Sociobiol.* 14 (1993): 209–229.

18. D. Daniels, "The Evolution of Concealed Ovulation and Self-Deception," *Ethology and Sociobiol.* 4 (1983): 69–87; P. W. Turke, "Effects of Ovulatory Concealment and Synchrony on Protohominid Mating Systems and Parental Roles," *Ethology and Sociobiol.* 5 (1984): 33–44. For a more cautionary view, see also H. D. Steklis and C. H. Whiteman, "Loss of Estrus in Human Evolution: Too Many Answers, Too Few Questions," *Ethology and Sociobiol.* 10 (1989): 417–434; I. Schroder, "Concealed Ovulation and Clandestine Copulation: A Female Contribution to Human Evolution," *Ethology and Sociobiol.* 14 (1993): 381–389.

19. R. M. Nesse and G. C. Williams, *Why We Get Sick* (New York: Random House, 1994).

20. M. Eals and I. Silverman, "The Hunter-Gatherer Theory of Spatial Sex

Differences: Proximate Factors Mediating the Female Advantage in Recall of Object Arrays," *Ethology and Sociobiol.* 15 (1994): 95–105.

21. W. D. Hamilton, "The Genetical Evolution of Social Behavior, 1, 2," *J. Theoretical Biology* 7 (1964): 1–15, 17–52; R. L. Trivers, "Parent-Offspring Conflict," *Amer. Zool.* 14 (1974): 249–264. For a recent review, see H. C. J. Godfray, "Evolutionary Theory of Parent-Offspring Conflict," *Nature* 376 (1995): 133–138.

22. A. Zahavi, "Reliability in Communication Systems and the Evolution of Altruism," in B. Stonehouse and C. Perrins, eds., *Evolutionary Ecology* (Baltimore, Md.: University Park Press, 1977), pp. 253–259.

23. D. Haig, "Genetic Conflicts in Human Pregnancy," *Quarterly Rev. Biol.* 68 (1993): 495–532.

24. For a recent example, see A. M. Warnecke, R. D. Masters, and G. Kempter, "The Roots of Nationalism: Nonverbal Behavior and Xenophobia," *Ethology and Sociobiol.* 13 (1992): 267–282. See also B. J. Craige, *American Patriotism in a Global Society* (Albany: State University of New York Press, 1996). One thesis of this book is that in the modern world, human predilections for tribalism promote nationalism, which in turn leads to a political dualism wherein love of country is linked to hatred of others. Another thesis is that such dualism is increasingly inappropriate in a global society.

25. From Darwin's (1871) *The Descent of Man,* as quoted in Craige, *American Patriotism.*

26. According to "Hamilton's rule" (Hamilton, "Genetical Evolution of Social Behavior"), a behavior is favored by selection whenever $\Delta w_x + \Sigma r_{yx} \Delta w_y > 0$, where Δw_x is the change the behavior causes in the individual's fitness, Δw_y is the change the behavior causes in the relative's fitness, and r_{yx} is the genetic relatedness of the individuals (e.g., $r = 0.5$ for full-sibs, $r = 0.25$ for half-sibs, and $r = 0.125$ for first cousins).

27. Another sociobiological possibility is that homosexuality in human ancestors served to minimize intragroup conflict and promote social harmony. In one of human's closest evolutionary relatives, the bonobo or pygmy chimpanzee *(Pan paniscus),* same-sex trysts are extremely common and serve to resolve power issues peacefully. See F. de Waal, *Bonobo: The Forgotten Ape* (Berkeley: University of California Press, 1997). On the other hand, if gay and lesbian lifestyles in humans were unknown to science, sociobiologists could glibly explain their absence as an expected consequence of the diminution in personal reproductive fitness.

28. E. O. Wilson, *On Human Nature* (Cambridge, Mass.: Harvard University Press, 1978).

29. R. Thornhill and N. W. Thornhill, "Human Rape: An Evolutionary Analysis," *Ethology and Sociobiol.* 4 (1983): 137–173.

30. This possibility has been considered in the sociobiological literature, and one

interpretation is that inbreeding is regulated societally not to avoid production of genetically defective children in extremely close matings, but instead to discourage more distant (e.g., cousin) matings that would serve to concentrate wealth and power within families, and thereby threaten the established rulers and lawmakers of society. See N. W. Thornhill, "The Evolutionary Significance of Incest Rules," *Ethology and Sociobiol.* 11 (1990): 113–129. For further discussion, see also C. V. J. Welham, "Incest, an Evolutionary Model," *Ethology and Sociobiol.* 11 (1990): 97–111.

31. M. B. Mulder, "Human Behavioral Ecology," in J. R. Krebs and N. B. Davies, eds., *Behavioral Ecology,* 3rd ed. (London: Blackwell, 1991), pp. 69–98.

32. Many books illustrate the extensive speculation that often accompanies scenarios about the selective agents in human evolution, but a fine recent example is K. Glantz and J. Pearce, *Exiles from Eden: Psychotherapy from an Evolutionary Perspective* (New York: Norton, 1989). Such "Pleistocentric" treatments often tend to view modern humans as genetically out of step with the modern world, or as "stone-agers in the fast lane."

33. D. S. Wilson, "Adaptive Genetic Variation and Human Evolutionary Psychology," *Ethology and Sociobiol.* 15 (1994): 219–235.

34. For general information, see E. O. Wilson, *Sociobiology, The New Synthesis* (Cambridge, Mass.: Harvard University Press, 1975). For descriptions of eusocial ants and wasps, see E. O. Wilson, *The Insect Societies* (Cambridge, Mass.: Harvard University Press, 1971). Eusociality refers to a suite of behaviors displayed by certain colony-living species, including cooperation in the care of young and reproductive divisions of labor with more or less sterile individuals working on behalf of the reproductives. For descriptions of matriphagy in spiders, see T. A. Evans, E. J. Wallis, and M. A. Elgar, "Making a Meal of Mom," *Nature* 376 (1995): 299. For more information on parthenogenesis, see R. M. Dawley and J. P. Bogart, eds., *Evolution and Ecology of Unisexual Vertebrates* (Albany: New York State Museum, 1989). Parthenogenesis or virgin birth is the development of an individual from a female gamete without the involvement of sperm.

35. C. J. Lumsden and E. O. Wilson, *Genes, Mind, and Culture* (Cambridge, Mass.: Harvard University Press, 1981); R. Dawkins, *The Selfish Gene* (New York: Oxford University Press, 1976).

36. A relatively new branch of mathematical population genetics attempts to describe quantitatively the theory of gene–culture coevolution. See M. W. Feldman and K. N. Laland, "Gene-Culture Coevolutionary Theory," *Trends Ecol. Evol.* 11 (1996): 453–457. In formal gene-culture models, individuals are described in terms of their "phenogenotype," the genotypic and phenotypic aspects of which are specified according to rules of Mendelian heredity and cultural transmission, respectively. Lactose absorption provides a straightforward example of a feature influenced by gene-culture coevolution. Individual humans are either lactose absorbers or malabsorbers, with absorption probably inherited as an autosomal dominant trait.

Human societies with long as opposed to short traditions of dairy farming tend to show much higher frequencies of lactose-absorption alleles. Formal gene-culture coevolutionary models attempt to account for such observations by examining the dynamics of allelic frequency change under various selection regimes, competing assumptions about the genetic bases of lactose absorption, and alternative modes of cultural transmission of milk usage.

37. For an introduction to this perspective, see papers in J. H. Barkow, L. Cosmides, and J. Tooby, eds., *The Adapted Mind: Evolutionary Psychology and the Generation of Culture* (New York: Oxford University Press, 1992).

38. Nesse and Williams, *Why We Get Sick,* provide an overview, and the following gives an example of a more specific hypothesis: J. R. Feierman, "A Testable Hypothesis about Schizophrenia Generated by Evolutionary Theory," *Ethology and Sociobiol.* 15 (1994): 263–282.

39. G. F. Eden, J. W. VanMeter, J. M. Rumsey, J. M. Maisog, R. P. Woods, and T. A. Zeffiro, "Abnormal Processing of Visual Motion in Dyslexia Revealed by Functional Brain Imaging," *Nature* 382 (1996): 66–69; C. Frith and U. Frith, "A Biological Marker for Dyslexia," *Nature* 382 (1996): 19–20.

40. "Reductionists Lay Claim to the Mind," *Nature* 381 (1996): 97.

41. S. Rose, "The Rise of Neurogenetic Determinism," *Nature* 373 (1995): 380–382.

42. S. Baron-Cohen, *Mindblindness: An Essay on Autism and Theory of Mind* (Cambridge, Mass.: MIT Press, 1995).

43. O. Sacks, *An Anthropologist on Mars: Seven Paradoxical Tales* (New York: Knopf, 1995). Oliver Sacks relates the remarkable story of an autistic child, Temple Grandin, who grew up to teach agricultural science at Colorado State University. Through amazing exercises of will, Grandin learned life's survival skills, yet she remains unable to appreciate concepts such as romantic love or empathy, or to develop meaningful emotional connections to others.

44. J. B. Martin, "Molecular Genetics of Neurological Diseases," *Science* 262 (1993): 674–676.

45. L. L. Hall, ed., *Genetics and Mental Illness* (New York: Plenum, 1996).

46. R. P. Ebstein et al., "Dopamine D4 Receptor (*D4DR*) Exon III Polymorphism Associated with the Human Personality Trait of Novelty Seeking," *Nature Genetics* 12 (1996): 78–80; J. Benjamin, B. Greenberg, D. L. Murphy, L. Lin, C. Patterson, and D. H. Hamer, "Population and Familial Association Between the D4 Dopamine Receptor Gene and Measures of Novelty Seeking," *Nature Genetics* 12 (1996): 81–84.

47. K.-P. Lesch et al., "Association of Anxiety-Related Traits with a Polymorphism in the Serotonin Transporter Gene Regulatory Region," *Science* 274 (1996): 1527–1531.

48. F. E. Bloom and D. J. Kupfer, eds., *Psychopharmacology: The Fourth Generation of Progress* (New York: Raven Press, 1995).

49. D. Johnson, "Can Psychology Ever Be the Same Again after the Genome is Mapped?" *Psychological Sci.* 1 (1990): 331–332.

50. F. Crick, *An Astonishing Hypothesis: The Scientific Search for the Soul* (New York: Scribner, 1994).

51. N. G. Waller, B. A. Kojetin, T. J. Bouchard, Jr., D. T. Lykken, and A. Tellegen, "Genetic and Environmental Influences on Religious Interests, Attitudes, and Values: A Study of Twins Reared Apart and Together," *Psychological Sci.* 1 (1990): 138–142.

52. Another conceivable form of personal empowerment through religion has been suggested by Herbert Benson in *Timeless Healing: The Power and Biology of Belief* (New York: Simon & Schuster, 1996), who claims that a powerful belief in a god manifests itself in placebo effects that improve an individual's bodily health. For critical reviews of this scientifically controversial perspective, see I. Tessman and J. Tessman, "Mind and Body," *Science* 276 (1997): 369–370; W. Roush, "Herbert Benson: Mind-Body Maverick Pushes the Envelope," *Science* 276 (1997): 357–359.

53. The thesis that the origins of human morality lie in biology is far from new, having been discussed by numerous writers. For a history and compilation of such thought, see R. D. Alexander, *The Biology of Moral Systems* (New York: Aldine De Gruyter, 1987). Not all biologists agree, however. For example, the prominent evolutionary biologist and Christian ideologist David Lack maintained that morality was of divine source: e.g., D. L. Lack, *Evolutionary Theory and Christian Belief; The Unresolved Conflict* (London: Methuen, 1957).

54. F. de Waal, *Good Natured: The Origins of Right and Wrong in Humans and Other Animals* (Cambridge, Mass.: Harvard University Press, 1996).

55. F. J. Ayala, "The Myth of Ethical Genes," *Trends Ecol. Evol.* 10 (1995): 470–471.

7. New Lords of Our Genes?

1. I recommend this book as an introduction to the scientific history of genetic engineering (New York: Norton, 1995).

2. Not all infectious plagues are behind us. Outbreaks such as AIDS, various influenzas, Legionnaires' disease, toxic shock syndrome, and Lyme disease provide powerful reminders that modern societies remain subject to major disease epidemics.

3. For example, the restriction enzyme *Eco*RI, named after the bacterium *Escherichia coli* from which it was isolated, cleaves duplex DNA wherever the (nonmethylated) nucleotide sequence GAATTC appears, whereas the restriction enzyme *Bam*H1 from *Bacillus ambofaciens* similarly cleaves at GGATCC.

4. S. Aldridge, *The Thread of Life* (Cambridge: Cambridge University Press, 1996).

5. For example, in eukaryotic cells short sugar molecules often are attached to proteins, generating glycoproteins that play an important role in chemical signaling necessary for proper cellular function.

6. Retroviruses, adenoviruses, herpes viruses, and other viral and nonviral delivery systems for transgenes are described in detail in D. L. Sokol and A. M. Gewirtz, "Gene Therapy: Basic Concepts and Recent Advances," *Critical Revs. Eukaryotic Gene Express.* 6 (1996): 29–57.

7. See, for example, T. M. Klein, L. Kornstein, J. C. Sanford, and M. E. Fromm, "Genetic Transformation of Maize Cells by Particle Bombardment," *Plant Physiol.* 91 (1989): 440–444.

8. M. D. Adams et al., "Initial Assessment of Human Gene Diversity and Expression Patterns Based upon 83 Million Nucleotides of cDNA Sequence," *Nature* 377S (1995): 3–174; G. D. Schuler et al., "A Gene Map of the Human Genome," *Science* 274 (1996): 540–546.

9. A. M. Maxam and W. Gilbert, "A New Method for Sequencing DNA," *Proc. Natl. Acad. Sci. USA* 74 (1977): 560–564; F. Sanger, S. Nicklen, and A. R. Coulson, "DNA Sequencing with Chain-Terminating Inhibitors," *Proc. Natl. Acad. Sci. USA* 74 (1977): 5463–5467.

10. The Human Genome Project began in 1989, and in the United States is administered by the National Institutes of Health and the Department of Energy with annual budgets to this point exceeding $200 million. The project's history is described in R. M. Cook-Deegan, *The Gene Wars: Science, Politics, and the Human Genome* (New York: Norton, 1994).

11. P. L. Ivanov, M. J. Wadhams, R. K. Roby, M. M. Holland, V. W. Weedn, and T. J. Parsons, "Mitochondrial DNA Sequence Heteroplasmy in the Grand Duke of Russia Georgij Romanov Establishes the Authenticity of the Remains of Tsar Nicholas II," *Nature Genetics* 12 (1996): 417–420; M. Krings, A. Stone, R. W. Schmitz, H. Krainitzki, M. Stoneking, and S. Pääbo, "Neanderthal DNA Sequences and the Origin of Modern Humans," *Cell* 90 (1997): 19–30; S. Pääbo, "Molecular Cloning of Ancient Mummy DNA," *Nature* 314 (1985): 644–645; G. H. Doran, D. N. Nickel, W. E. Ballinger, Jr., O. F. Agee, P. J. Laipis, and W. W. Hauswirth, "Anatomical, Cellular and Molecular Analysis of 8,000-Yr-Old Human Brain Tissue from the Windover Archaeological Site," *Nature* 323 (1996): 803–806. Actually, some of these studies used traditional pre-PCR methods for DNA recovery. The development of PCR technology makes such tasks simpler and more powerful, such that DNA sequences now are obtained almost routinely from various ancient human remains.

12. Mini- and microsatellite regions of the genome conventionally are referred

to as genes or loci, although in many cases they do not encode functional protein products and their operational functions if any within cells remain unclear.

13. Electrophoretic separations lie at the heart of many other DNA- and protein-level assays as well. For example, DNA sequencing involves separations of DNA molecules that differ in length by as little as one base pair.

14. E. M. Southern, "Detection of Specific Sequences among DNA Fragments Separated by Gel Electrophoresis," *J. Molec. Biol.* 98 (1975): 503–517. Related molecular assays developed later were dubbed Northern blotting (used to identify RNAs) and Western blotting (used to identify specific proteins).

15. A. Abbott, "DNA Chips Intensify the Sequence Search," *Nature* 379 (1996): 392.

16. The ICSI technique more typically has been employed for men with low sperm counts, using their rare but mature spermatozoa rather than immature spermatids. See D. Butler, "Spermatid Injection Fertilizes Ethics Debate," *Nature* 377 (1995): 277; R. J. Aitken and D. S. Irvine, "Fertilization Without Sperm," *Nature* 379 (1996): 493–495. At least two potential dangers attend ICSI techniques. First, if the infertility is based on genes and heritable, as is true in many cases, some offspring produced by the ISCI procedure might themselves be sterile [see J. L. Pryor et al., "Microdeletions in the Y Chromosome of Infertile Men," *New England J. Medicine* 336 (1997): 534]. Second, the infertility syndrome can be associated with other heritable disorders, thus further jeopardizing the health of ICSI babies.

17. One related approach involves the use of hormones to trigger a woman's ovulation prior to IVF. Another is gamete intrafallopian transfer (GIFT), in which a doctor uses a laparoscope to insert eggs and sperm directly into a woman's fallopian tube, after which any resulting embryos may travel naturally to her uterus. Still another is zygote intrafallopian transfer (ZIFT), in which a zygote formed by the union of egg and sperm in a test tube is inserted into a woman's fallopian tube.

18. In a new method called oocyte farming, applied thus far only to mice, eggs can be cultivated *in vitro* all the way from primordial precursor cells to mature, fertilizable gametes. This revolutionary technique promises to increase greatly the yield of eggs that can be harvested for IVF or for other research purposes. See W. Roush, "Fertile Results: Bringing up Baby (Eggs)," *Science* 271 (1996): 594–595.

19. I. Wilmut, A. E. Schnieke, J. McWhir, A. J. Kind, and K. H. S. Campbell, "Viable Offspring Derived from Fetal and Adult Mammalian Cells," *Nature* 385 (1997): 810–813. Technically, the lamb may not have been strictly clonal because, although its nuclear DNA was identical to that of its "clonemate" mother, some of its mitochondrial DNA probably came from the "surrogate" mom.

20. The kit merely provides a simple way for people to procure a sample of their own DNA suitable for HIV testing. Samples then are mailed to a clinical lab where the tests are conducted.

21. Amniocentesis involves insertion of a hollow needle through a mother's abdomen and into the amniotic sac that surrounds the fetus. There, cells naturally sloughed off from the fetus can be extracted through the needle. In the CVS procedure, a catheter is introduced through the mother's vagina, and cells are sampled from the outer envelope (chorion) of the embryo.

22. The various preimplantation genetic procedures described here are still only at an experimental stage of development. In preliminary trials, they have been employed to screen for the AAT deficiency gene in eggs cells, and for the cystic fibrosis gene in early blastocysts.

23. This sentiment was attributed to Dr. Neil Holtzman of Johns Hopkins University, in the book *Altered Fates*. On the other hand, technological developments may improve this situation greatly. One promising method under development involves the use of robot-constructed oligonucleotide microchips to increase the number of patients who can be screened for genetic disorders such as β-thalassemia. See G. Yershov et al., "DNA Analysis and Diagnostics on Oligonucleotide Microchips," *Proc. Natl. Acad. Sci. USA* 93 (1996): 4913–4918.

24. W. F. Anderson, "Gene Therapy," *Sci. Amer.* 274 (1995): 124–128.

25. Some approaches to extending the *in vivo* "shelf-life" of injected proteins are available. In a PEG-ADA procedure, for example, the enzyme adenosine deaminase is coated with a waxy, protective polymer called polyethylene glycol before it is injected into a patient.

26. Not all genetic therapies will imbue a patient with permanent capabilities for *in vivo* production of the required protein: The outcome depends on additional factors including the types of cells genetically transformed. Mature blood cells, for example, have limited life spans, so gene therapy treatments targeted to them must be reapplied periodically. Genes transferred to bone marrow stem cells (the immature precursors of blood cells) should have longer effectiveness because these cells essentially are immortal within an individual.

27. The two principal bodies that oversee the approval of human gene therapy protocols are the Recombinant DNA Advisory Committee (RAC) of the National Institutes of Health, and the Food and Drug Administration (FDA).

28. G. H. Gibbons and V. J. Dzau, "Molecular Therapies for Vascular Diseases," *Science* 272 (1996): 689–693.

29. For example, familial hypercholesterolemia (FH) is a rare genetic disorder, inherited as a dominant mutation in a defective receptor gene, that gives its bearers high cholesterol readings and a strong likelihood of heart attacks beginning between the ages of thirty and forty. In 1992, a patient suffering from FH was the subject of an experimental procedure in which some of her liver cells were removed surgically, transduced with good copies of the receptor gene, and returned to her liver via catheter. The outcome of this procedure was unclear: Although her cholesterol levels

subsequently dropped by about 25 percent, the progression of atherosclerosis appeared not to be arrested.

30. A recent panel report for the National Institutes of Health (which spends about $200 million a year on gene therapy research) gave a tough review of some gene therapists and their sponsors for overselling the technology and its achievements to the present time. See E. Marshall, "Gene therapy's growing pains," *Science* 269 (1995): 1050–1055; E. Marshall, "Less Hype, More Biology Needed for Gene Therapy," *Science* 270 (1995): 1751. On the other hand, proponents of gene therapy point out that to unduly criticize this promising young field would be premature [see a recent section of *Scientific American* devoted to gene therapy: 276 (1997): 95–123].

31. For one of the first discussions of moral problems in genetics, see J. Glover, *What Sort of People Should There Be?* (New York: Penguin, 1984). Examples of recent books on this general subject include K. A. Drlica, *Double-Edged Sword: The Risks and Promises of the Genetic Revolution* (Reading, Mass.: Addison-Wesley, 1994); S. Jones, *The Language of Genes: Solving the Mysteries of Our Genetic Past, Present and Future* (New York: Doubleday, 1994); S. Jones, *In the Blood: God, Genes and Destiny* (New York: Harper-Collins, 1996); P. Kitcher, *The Lives to Come: The Genetic Revolution and Human Possibilities* (New York: Simon & Schuster, 1996).

32. Assessments of the potential economic and societal impact of the biotechnology industry can be found in the following: U.S. Congress, Office of Technology Assessment, *Biotechnology in a Global Economy,* OTA-BA-494 (Washington, D.C.: U.S. Government Printing Office, 1991); National Research Council, *Putting Biotechnology to Work* (Washington, D.C.: National Academy Press, 1992).

33. Useful discussions of the history and controversy surrounding gene patent issues may be found in the following: C. T. Caskey, "Gene Patents—A Time to Balance Access and Incentives," *Trends Biotechnol.* 14 (1996): 298–302; R. S. Eisenberg, "Intellectual Property Issues in Genomics," *Trends Biotechnol.* 14 (1996): 302–307. See also E. Marshall, "A Showdown over Gene Fragments," *Science* 266 (1994): 208–210; D. Dickson, "Open Access to Sequence Data 'Will Boost Hunt for Breast Cancer Gene,'" *Nature* 378 (1995): 425; D. Dickson, "'Leak' Rumours Fuel Debate on Gene Patent," *Nature* 379 (1996): 574.

34. See S. M. Thomas, A. R. W. Davies, N. J. Birtwistle, S. M. Crowther, and J. F. Burke, "Ownership of the Human Genome," *Nature* 380 (1996): 387–388.

35. D. Dickson, "Whose Genes are They Anyway?" *Nature* 381 (1996): 11–14.

36. The statement was issued by the Foundation for Economic Trends and General Board of Church and Society of the United Methodist Church.

37. R. Cole-Turner, "Religion and Gene Patenting," *Science* 270 (1995): 52. The author is a professor of theology at the Memphis Theological Seminary and a member of the American Association for the Advancement of Science Committee

on Scientific Freedom and Responsibility. See also R. Cole-Turner, *The New Genesis: Theology and the Genetic Revolution* (Louisville, Ky: Westminster John Knox, 1993).

38. See also B. Cohen, "Population Groups Can Hold Critical Clues," *Nature* 381 (1996): 12.

39. Prospecting for human genes is one matter, but similar concerns about possible exploitation apply as well to the commercialization of plant and animal gene products, for example by international drug companies hunting for new chemicals in third-world countries.

40. E. Masood, "Gene Tests: Who Benefits from The Risk?" *Nature* 379 (1996): 389–392.

41. Egg and sperm cells are unique living forms too, with an existence definable as originating at the completion of meiosis. Pro-lifers haven't yet demanded legal rights for gametes.

42. The following is an "Instruction on Respect for Human Life" *(Donum Vitae)* from the teaching office of the Catholic church: "the fruit of human generation from the first moment of its existence, that is to say, from the moment the zygote has formed, demands the unconditional respect that is morally due to the human being in his bodily and spiritual totality. The human being is to be respected and treated as a person from the moment of conception and therefore from that same moment his rights as a person must be recognized, among which in the first place is the inviolable right of every innocent human being to life." Other moral positions are less extreme. For example, the Ethics Committee of the American Fertility Society concluded "proximity to birth is a principal factor in apportioning greater moral value on the developing human life" ["Ethical Considerations of the New Reproductive Technologies," *Fertility and Sterility* 46 (1986): supplement 1].

43. An early advocate for such a committee was James Watson, one of the codiscoverers of the double-helical structure of DNA and the initial director of the Human Genome Project. The first chairperson named to head the ELSI was Nancy Wexler, a humanist pioneer in the history of research on genetic disease.

44. See, for example L. S. Parker and E. Gettig, "Ethical Issues in Genetic Screening and Testing, Gene Therapy, and Scientific Conduct," in F. E. Bloom and D. J. Kupfer, eds., *Psychopharmacology: The Fourth Generation of Progress* (New York: Raven Press, 1995), pp. 1875–1881.

8. Meaning

1. S. Weinberg, *Dreams of a Final Theory* (New York: Vintage Books, 1994).

2. From number 230 in *Pensees,* a collection of Pascal's unfinished writings published after his death at age thirty-nine.

3. Ibid., number 895.

4. *The Brothers Karamazov,* trans. C. Garnett (New York: Random House, 1933).

5. E. O. Wilson, *On Human Nature* (Cambridge, Mass.: Harvard University Press, 1978).

6. G. Vidal, "The Great Unmentionable—Monotheism and its Discontents," *Nation,* July 13, 1992, p. 37.

7. B. Russell, *Why I Am Not a Christian* (New York: Simon & Schuster, 1957).

8. W. B. Provine, "Evolution and Meaning in Life." Abstract in the symposium, *Current Views on the History of Organisms,* Kawasaki, Japan, 1993.

9. There are many biographies of Darwin. For a close personal portrait, see J. Brown, *Charles Darwin: Voyaging* (Princeton, N.J.: Princeton University Press, 1995).

10. See *Science, Philosophy, and Religion, A Symposium.* Conference on Science, Philosophy, and Religion in Their Relation to the Democratic Way of Life, New York, 1941.

11. However, other famous physicists have held varying opinions on the existence of a god. Isaac Newton believed that God made the Scriptures as well as the book of nature for human scholars to study, and that both were equally real. Stephen Hawking's scientific theory, that the universe is closed and has no beginning or end, is a view that could be interpreted to challenge the traditional theological stance that a god must at some point have initiated creation even if he no longer assumes an active role. See S. Hawking, *A Brief History of Time* (New York: Bantam Books, 1988).

12. I find it intellectually unburdening and emotionally uplifting to accept that humans arose from natural forces rather than through a supernatural god. However, such feelings (or those of conventional theologists) are not valid criteria for deciding between competing hypotheses on human origins.

13. C. M. Turnbull, *The Mountain People* (New York: Simon & Schuster, 1972).

14. "Mental adjustment" might be mentioned as well, but for present purposes this would be defined as an epiphenomenon influenced by genes and environment.

15. Although this would be difficult to prove, I suggest that this is true for other species as well. Certainly, mobile organisms tend actively to seek out environments conducive to their own survival and reproduction.

16. Although precise definitions of tool employment are difficult, scores of reasonable examples exist for other species. These range from the collection and use of shell decorations by masking crabs, to elaborate nest adornments by many birds, to the construction and use of specialized dipping sticks by chimpanzees to extricate termites (a food delicacy) from subterranean nests See B. B. Beck, *Animal Tool Behavior: The Use and Manufacture of Tools by Animals* (New York: Garland STPM Press, 1980).

17. B. Campbell, *Human Evolution: An Introduction to Man's Adaptations* (Chicago: Aldine, 1966).

18. See P. R. Ehrlich, *The Population Bomb* (New York: Ballentine Books, 1968).

19. National Academies of Science, New Delhi, India.

20. A notable exception is a recent book by Vice President Al Gore in which he demonstrates exceptional insight into environmental issues and human ecology. See *Earth in the Balance* (New York: Penguin Books, 1992).

21. Almost no stands of virgin forest remain today in the continental United States. In their travel diaries, naturalists such as William Bartram and John Muir described magnificent primary forests that bear little resemblance to the second-growth remnants in today's urban areas and farmlands.

22. In the United States, about four million miles of road, utilized by 200 million vehicles, cover about 2 percent of the country's land surface.

23. Until recently little attention was paid to such topics as "landscape architecture" and "environmental design." Only in recent years have substantial numbers of universities implemented such curricula, or have environmental consulting firms made it their business.

24. A recent "biophelia hypothesis" proposes that humans have innate emotional affiliations to other organisms such that our psychological well-being is enhanced by exposure to biodiversity. See E. O. Wilson, *Biophilia* (Cambridge, Mass.: Harvard University Press, 1984), and S. R. Kellert and E. O. Wilson, eds., *The Biophilia Hypothesis* (Washington, D.C.: Island Press, 1993).

25. D. M. Byers, "Religion and Science: The Emerging Dialogue," *America* 174 (1996): 8–15.

26. "Science and the Creation of Life," *Origins* 17 (1987): 23.

27. Among the Church's publications resulting from this ongoing dialogue are *Science and the Catholic Church* (U.S. Catholic Conference Publishing Office, 1995), a brochure declaring a unilateral withdrawal from the war between religion and science, and *Critical Decisions: Genetic Testing and Its Implications* (U.S. Catholic Conference Publishing Office, 1996).

28. The pope, however, did not relinquish to evolution's jurisdiction the human soul, asserting instead that the soul is created anew, divinely, in each person. There also has been some controversy about the precise wording as well as the meaning of the pope's translated statement.

29. This includes the National Council of Churches (Protestants), U.S. Catholics Conference, and the Coalition on the Environment and Jewish Life.

Glossary

adaptation	Any feature (morphological, physiological, behavioral) that makes an organism better suited to survive and reproduce in a particular environment.
adenine	One of the four organic bases normally composing DNA.
aging	See senescence.
allele	Any of the possible alternative forms of a given gene. A diploid individual carries two alleles at each autosomal gene, and these can either be identical in state (in which case the individual is homozygous) or different in state (heterozygous). At each autosomal gene, a population of N diploid individuals harbors $2N$ alleles, many of which may differ in details of nucleotide sequence.
altruism	Self-harmful behavior performed for the benefit of others.
amino acid	One of the molecular subunits polymerized to form polypeptides.
amniocentesis	A clinical procedure for prenatal genetic diagnosis in which a hollow needle inserted through the mother's abdomen is used to obtain fetal cells sloughed into the amniotic fluid.
antibody	A protein produced by cells of the immune system that binds specifically to a foreign substance and initiates the immune response.
antigen	A foreign substance that upon introduction to a vertebrate animal stimulates the production of antibodies.
apoptosis	Genetically programmed cell death, often a normal feature of organismal development and differentiation.

261

artificial selection — Selection that operates in plant or animal populations through the human choice of particular genetically-based traits.

asexual reproduction — Any form of reproduction that does not involve the fusion of sex cells (gametes).

autoradiograph — A photograph of the position of radioactively labeled substances, as for example in a cell or in an electrophoretic gel.

autosome — A chromosome in the nucleus other than a sex chromosome; in diploid organisms, autosomes are present in homologous pairs. See also sex chromosome.

bacterium — A unicellular microorganism without a true cellular nucleus.

blastocyst — A mammalian embryo at the time of its implantation into the uterine wall.

blastomere — One of the cells into which the egg divides during initial cleavages.

blastula — A hollow sphere of cells resulting from early cell cleavages during embryonic development.

cancer — A disease characterized by uncontrolled cellular proliferation.

cell — A small, membrane bound unit of life capable of self-reproduction.

cell theory — The idea put forth by Matthias Jacob Schleiden and Theodor Schwann in 1839 that all life is composed of cells, and that organismal growth and reproduction result from cell divisions.

chimera — An individual composed of a mixture of genetically different cells. The word sometimes is restricted to apply to such cells when they derive from separate zygotes. See also mosaic.

chromosome — A threadlike structure within a cell that carries genes.

clone — A group of genetically identical cells or organisms, all descended from a single ancestral cell or organism, or the process of creating such genetically identical cells or organisms.

coalescent theory The body of mathematics and thought concerning how alleles in organisms alive today trace back through population pedigrees to common ancestral states.

codominance A genetic situation in which both alleles in a heterozygous diploid individual are expressed simultaneously in the phenotype.

complementary DNA (cDNA) DNA produced from a RNA template by a reversal of the transcription process.

cytoplasm The portion of a eukaryotic cell outside of the nucleus.

cytosine One of the four organic bases normally composing DNA.

deletion The loss of a segment of genetic material from a chromosome.

deoxyribonucleic acid (DNA) A double-stranded polynucleotide molecule each of whose nucleotide subunits is composed of a deoxyribose sugar, a phosphate group, and one of the nitrogenous bases adenine, guanine, cytosine, or thymine.

diploid A usual condition of a somatic cell in which two copies of each chromosome are present. See also haploid.

dizygotic twins Genetically nonidentical siblings that stem from two separate zygotes during a pregnancy (fraternal twins). See also monozygotic twins.

DNA/DNA hybridization Any of a class of laboratory procedures in which single-strand stretches of polynucleotides attract and bind to homologous, complementary single strands.

DNA fingerprint The complex, individual pattern of genetic material as revealed on electrophoretic gels following separation of multiple DNA restriction fragments. See also minisatellite locus and microsatellite locus.

DNA ligase An enzyme that catalyzes the end-to-end molecular union of DNA fragments.

dominance A genetic situation in which one allele in a heterozygous diploid individual is expressed fully in the phenotype. See also recessivity.

duplication   A genetic event usually stemming from an abnormal meiosis in which a gene or portion of a chromosome gives rise by any of several cellular mechanisms to a second copy.

electrophoresis   The movement of charged molecules such as proteins or nucleic acids through a supporting medium (starch, agarose, or acrylamide gel) under the influence of an electric current.

embryo   An organism in the early stages of development (in humans, usually up to the beginning of the third month of pregnancy).

enzyme   A protein that catalyzes a specific chemical reaction.

epigenetics   The entire suite of mechanisms, developmental pathways, and social and other environmental influences by which genomes yield organismal-level features.

epistasis   A genetic situation in which two or more genes influence one another's expression in the production of a phenotype.

eugenics   The ideology or practice of attempting to improve *Homo sapiens* by altering its genetic composition.

eukaryote   Any organism in which chromosomes are housed in a true membrane-bound nucleus.

eusocial   A social system, such as of many bees, wasps, and ants, characterized by cooperative care of young and reproductive division of labor with sterile individuals working on behalf of reproducers within a colony.

evolution   Change through time in the genetic composition of populations.

exon   A coding segment of a gene. See also intron.

expressed sequence tag (EST)   A portion of the coding sequence of a gene identified from its messenger RNA.

fetus   An organism in intermediate stages of development in the uterus (in humans, beginning at about the third month of pregnancy).

fitness (genetic) The contribution of a genotype to the next generation relative to the contributions of other genotypes in the population.

fossil Any remain or trace of life no longer alive.

gamete A mature reproductive sex cell (egg or sperm).

gametogenesis The specialized series of cellular divisions that leads to the production of gametes. See also meiosis.

gastrula A two-layered, cup-shaped stage of embryonic development.

gene The basic unit of heredity; usually taken to imply a sequence of nucleotides specifying production of a polypeptide or other functional product such as ribosomal RNA, but also can be applied to stretches of DNA with unknown or unspecified function.

genealogy A record of descent from ancestors through a pedigree.

gene pool The sum total of all hereditary material in a population or species.

gene therapy The insertion of a functional gene into an individual's cells with the intent of correcting a hereditary disorder.

genetic drift Change in allele frequency in a finite population by chance sampling of gametes from generation to generation.

genetic engineering Any experimental or industrial method employed to alter the genomes of living cells.

genetic load The collective burden of genetic defects in a population.

genocopy A genetic-based phenotypic condition that mimics an effect normally induced by the environment.

genome The complete genetic constitution of an organism; also can refer to a particular composite piece of DNA, such as the mitochondrial genome.

genotype The genetic constitution of an individual organism with reference to a single gene or set of genes. See also phenotype.

germ cell

A sex cell or gamete. See also somatic cell.

group selection

Selection that operates on multiple members of a hereditary lineage as a unit.

guanine

One of the four organic bases normally composing DNA.

haploid

A usual condition of a gametic cell in which one copy of each chromosome is carried. See also diploid.

hemizygous

A gene present in a single dose, such as a sex-linked gene in the heterogametic sex.

heredity

Inheritance of genes; the phenomenon of familial transmission of genetic material from one generation to the next.

heritability

The fraction of variation of a trait within a population due to heredity as opposed to environmental influences.

hermaphrodite

A condition in which an individual displays both testicular and ovarian development.

heterogametic sex

The gender that produces gametes containing unlike sex chromosomes (in humans, the male). See also homogametic sex.

heteroplasmy

The co-occurrence of two or more different cytoplasmic genotypes (such as those for mtDNA) within a cell or individual.

heterosis

The condition in which heterozygotes have higher genetic fitness than homozygotes.

heterozygote

A diploid organism possessing two different alleles at a specified gene. See also homozygote.

homeobox

Specific DNA sequences that regulate patterns of morphological differentiation during organismal development.

homeotic gene

A gene that controls the overall body plan of an organism by influencing the developmental fate of groups of cells.

homogametic sex	The gender that produces gametes containing alike sex chromosomes (in humans, the female). See also heterogametic sex.
homology	Similarity of structure due to inheritance from a shared ancestor; can refer to any structural features ranging from DNA sequences to morphological traits.
homozygote	A diploid organism possessing two identical alleles at a specified gene. See also heterozygote.
hormone	An organic compound produced in one region of an organism and transported to target cells in other parts of the body where its effects on phenotype are exerted.
inbreeding	The mating of kin.
inclusive fitness	The sum of an individual's personal genetic fitness plus that individual's influences on genetic fitness in relatives other than direct descendants.
independent assortment (Mendel's law of)	The random distribution to gametes of the alleles from genes on different chromosomes, or from genes far enough apart on a given chromosome.
intron	A noncoding portion of a gene. Most genes in eukaryotes consist of alternating intron and exon DNA sequences.
inversion	A genetic condition in which a chromosomal segment has been rotated 180° from its original linear orientation.
jumping gene	See transposable element.
kin selection	A form of natural selection due to one or more individuals favoring or disfavoring the survival and reproduction of genetic relatives other than offspring.
Lamarckian inheritance	A theoretical mode of inheritance (apparently undocumented) wherein somatic features acquired during the lifetime of an individual are passed genetically to progeny.
locus (pl. loci)	A gene.
Lyon effect	The inactivation of one of the two X chromosomes in each somatic cell of a female mammal.

meiosis	The cellular process whereby a diploid cell divides to form haploid gametic cells.
meiotic drive	Any mechanism in meiosis that results in the unequal recovery of the two types of gametes produced by a heterozygote.
messenger RNA	A form of ribonucleic acid transcribed from structural genes, the exon-derived portions of which subsequently will be translated into a polypeptide.
metabolism	The sum of all physical and chemical processes by which living matter is produced and maintained, and by which cellular energy is made available.
microsatellite locus	A stretch of DNA containing short, repeated sequences each typically about two, three, or four base pairs in length.
minisatellite locus	A stretch of DNA containing repeated sequences each typically about twenty to a hundred base pairs in length.
mitochondrion	An organelle in the cell cytoplasm that contains its own DNA (mtDNA) and that is the site of some of the metabolic pathways involved in cellular energy production.
mitosis	A process of cell division that produces daughter cells with the same chromosomal constitution as the parent cell. See also meiosis.
mobile element	See transposable element.
monogamy	A mating system in which one male is paired with one female.
monomer	A simple compound from which, by repetition of a single reaction, a polymer is formed.
monozygotic twins	Genetically identical siblings (barring mutation) that stem from a single zygote during a pregnancy. See also dizygotic twins.
morphogen	A biochemical substance that provides positional information in the form of a concentration gradient influencing embryonic development.

mosaic	An individual composed of a mixture of genetically different cells derived from the same zygote. See also chimera.
Muller's ratchet	The tendency for deleterious mutations to irreversibly accumulate in a population completely lacking genetic recombination.
mutation	A change in the genetic constitution of an organism.
natural selection	The differential contribution by individuals of different genotypes to the population of offspring in the next generation.
nucleic acid	See deoxyribonucleic acid or ribonucleic acid.
nucleotide	A unit of DNA or RNA consisting of a nitrogenous base, a pentose sugar, and a phosphate group.
nucleus	The portion of a eukaryotic cell bounded by a nuclear membrane and containing chromosomes.
oligonucleotide	A polymer made up of a small number (two to thirty) of nucleotides.
oncogene	A gene that induces uncontrolled cell proliferation. See also cancer.
ontogeny	The development of an individual from fertilized egg to maturity.
oocyte	An egg cell.
outcrossing	The pairing of unrelated individuals.
overdominance	A genetic situation in which diploid individuals who are heterozygous at a particular gene have more extreme phenotypes than either homozygote.
parthenogenesis	The development of an individual from an egg without fertilization. See also asexual reproduction.
pattern formation	The emergence of specialized tissues and body parts in appropriate locations in the developing individual.
pedigree	A diagram displaying ancestry (mating partners and off-spring produced across generations) within a population.
phenocopy	An environmentally induced phenotypic condition that

mimics an effect normally caused by genetic constitution.

phenotype | The observable properties of an organism at any level, ranging from molecular and physiological to gross morphological.

phenotypic plasticity | The capacity of different phenotypes to emerge when a genotype is exposed to different environmental conditions.

phylogeny | Evolutionary relationships (historical descent) of a group of organisms or species.

plasmid | A small extra-chromosomal genetic element found in bacteria.

pleiotropy | A genetic phenomenon wherein a single gene influences multiple phenotypic features.

Pleistocene Epoch | The geological time frame beginning about 2 million years ago and ending roughly 10,000 years before the present.

polar body | The nonfunctional, haploid cellular products of meiosis other than the oocyte.

polyandry | A mating system in which a female acquires and mates with multiple males. See also polygyny and polygamy.

polygamy | A mating system in which an individual has more than one mate. See also polygyny and polyandry.

polygene | Any of the different genes that can affect the same phenotypic trait.

polygenic trait | A phenotypic trait affected by multiple genes.

polygyny | A mating system in which a male acquires and mates with multiple females. See also polyandry and polygamy.

polymer | A large molecule composed of a bonded collection of repeating subunits (monomers) linked together during a series of similar chemical reactions.

polymerase | An enzyme that catalyzes the formation of nucleic acid molecules.

polymerase chain reaction (PCR)	A laboratory procedure for the *in vitro* replication of DNA from small starting quantities; PCR involves repeated cycles of DNA denaturation, primer annealing, and primer extension.
polymerization	The formation of a polymer from a collection of monomeric molecules.
polymorphism	With respect to particular features, the presence of two or more genetic conditions or phenotypes in a population.
polypeptide	A polymer composed of amino acids chemically linked together.
population bottleneck	A severe but often temporary reduction in the size of a population.
primer (for PCR)	An oligonucleotide used in conjunction with a polymerase to initiate synthesis of a nucleic acid.
prokaryote	Any microorganism that lacks a chromosome-containing, membrane-bound nucleus.
protein	A macromolecule composed of one or more polypeptide chains.
protozoan	A unicellular animal.
pseudogene	A gene bearing close structural resemblance to a known functional gene at another chromosomal site, but that itself is nonfunctional due to genetic alterations such as additions, deletions, or nucleotide substitutions.
purine	Either of the organic bases adenine or guanine.
pyrimidine	Either of the organic bases thymine or cytosine.
recessivity	A genetic situation in which one allele in a heterozygous diploid individual is masked in phenotypic expression by a dominant allele at the same locus. See also dominance.
recombinant DNA	A new DNA molecule that has arisen from genetic recombination.
recombination (genetic)	The formation of new combinations of genes through such natural processes as meiosis and fertilization, or in the laboratory through recombinant DNA technologies.

regulatory gene	A gene that exerts operational control over the expression of other genes.
restriction enzyme	An enzyme produced by a bacterium that cleaves foreign DNA molecules at specific oligonucleotide recognition sites. Restriction enzymes are used widely in recombinant DNA technology.
restriction fragment	A linear segment of DNA resulting from cleavage of a longer segment by a restriction enzyme.
retrotransposable element	A mobile element that moves about the genome by an intermediate RNA molecule which then is reverse-transcribed into DNA.
retrotransposon	See retrotransposable element.
retrovirus	An RNA virus that utilizes reverse transcription during its life cycle to integrate into the DNA of host cells.
ribonucleic acid (RNA)	A single-stranded polynucleotide molecule each of whose nucleotide subunits is composed of a ribose sugar, a phosphate group, and one of the nitrogenous bases adenine, guanine, cytosine, or uracil.
ribosomal RNA	A form of ribonucleic acid that together with ribosomal proteins composes a ribosome.
ribosome	An organelle in the cell cytoplasm composed of RNA and protein that is the site of protein translation.
segregation (Mendel's law of)	The distribution to gametes of the two alleles in a diploid individual; each gamete receives, at random, one or the other of the two alleles at each gene.
selfish DNA	DNA that displays self-perpetuating modes of behavior without apparent benefit to the organism.
senescence	A persistent decline in the age-specific survival probability or reproductive output of an individual due to internal physiological deterioration.
sex chromosome	A chromosome in the cell nucleus involved in distinguishing the two genders. In humans, the "X" and "Y" are sex chromosomes. See also autosomes.

sexual reproduction	Reproduction involving the production and subsequent fusion of haploid gametes.
sexual selection	The differential ability of individuals of the two genders to acquire mates. Intrasexual selection refers to competition among members of the same gender over access to mates; intersexual selection refers to choices made between males and females.
sociobiology	The scientific study of the biological basis of social behaviors.
somatic cell	Any cell in a eukaryotic organism other than those destined to become germ cells. See also germ cell.
Southern blotting	A technique developed by E. M. Southern for transferring electrophoretically separated DNA fragments from an electrophoretic gel to a nitrocellulose filter for subsequent hybridization against a DNA probe. The method is used to identify particular genes or gene fragments from an unknown sample.
spermatid	One of the haploid cells produced by meiosis in males. A spermatid matures to a sperm cell.
syngamy	The union of two gametes to produce a zygote; fertilization.
thymine	One of the four organic bases normally composing DNA.
totipotency	The capacity retained by some cells to differentiate and proliferate into the diverse cell types of an adult organism.
transcription	The cellular process by which an RNA molecule is formed from a DNA template.
transfer RNA	A form of ribonucleic acid that picks up amino acids from the cell cytoplasm and moves them into position for the translation process.
transgene	A foreign gene carried by a transgenic organism.
transgenic organisms	Organisms containing injected genetic material from another organism or species.

translation	The process by which the genetic information in messenger RNA is employed by a cell to direct the construction of polypeptides.
translocation	An interchange of chromosomal segments between non-homologous chromosomes, or between distant regions of homologous chromosomes.
transposable element	Any of a class of DNA sequences that can move from one chromosomal site to another, often replicatively.
transposition	The process by which a replica of a transposable element is inserted into another chromosomal site.
uracil	An organic base in RNA that replaces thymine in the corresponding DNA.
virus	A tiny, obligate intracellular parasite incapable of autonomous replication, but which instead utilizes the host cell's replicative machinery.
X chromosome	The sex chromosome normally present as two copies in female mammals (the homogametic sex), but as only one copy in males (the heterogametic sex).
Y chromosome	In mammals, the sex chromosome normally present in males only.
zygote	The diploid cell arising from the union of male and female haploid gametes.

Index

275